S0-BIP-208

Guides to Information Sources

Information Sources in

Polymers and Plastics

Guides to Information Sources

A series under the General Editorship of
D. J. Foskett, MA, FLA
and
M. W. Hill, MA, BSc, MRIC

This series was known previously as 'Butterworths Guides to Information Sources'.

Other titles available are:

Information Sources in Energy Technology
edited by L. J. Anthony

Information Sources in the Life Sciences
edited by H. V. Wyatt

Information Sources in Physics (Second edition)
edited by Dennis F. Shaw

Information Sources in Law
edited by R. G. Logan

Information Sources in Management and Business (Second edition)
edited by K. D. C. Vernon

Information Sources in Politics and Political Science: a survey worldwide
edited by Dermot Englefield and Gavin Drewry

Information Sources in Engineering (Second edition)
edited by L. J. Anthony

Information Sources in Economics (Second edition)
edited by John Fletcher

Information Sources in the Medical Sciences (Third edition)
edited by L. T. Morton and S. Godbolt

Guides to Information Sources

Information Sources in
Polymers and Plastics

Editor
R. T. Adkins, MIInfSc
Consultant Information Scientist,
Hatfield, Hertfordshire

BOWKER-SAUR
London • Edinburgh • Munich • New York
Singapore • Sydney • Toronto • Wellington

British Library Cataloguing in Publication Data

Information sources in polymers and plastics.—(Guides to Information Sources)
1. Polymers. Information sources. 2. Plastics. Information sources
I. Adkins, Ronald II. Series
547.7'07

ISBN 0-408-02027-X

Bowker-Saur Ltd is part of the European Division of Butterworths, Borough Green, Sevenoaks, Kent TN15 8PH

Cover design by Calverts Press
Printed on acid-free paper
Printed and bound in Great Britain by
Biddles Ltd. Guildford and King's Lynn

Series Editors' Foreword

Daniel Bell has made it clear in his book *The Post-Industrial Society* that we now live in an age in which information has succeeded raw materials and energy as the primary commodity. We have also seen in recent years the growth of a new discipline, information science. This is in spite of the fact that skill in acquiring and using information has always been one of the distinguishing features of the educated person. As Dr Johnson observed, 'Knowledge is of two kinds. We know a subject ourselves, or we know where we can find information upon it.'

But a new problem faces the modern educated person. We now have an excess of information, and even an excess of sources of information. This is often called the 'information explosion', though it might be more accurately called the 'publication explosion'. Yet it is of a deeper nature than either. The totality of knowledge itself, let alone of theories and opinions about knowledge, seems to have increased to an unbelievable extent, so that the pieces one seeks in order to solve any problem appear to be but a relatively few small straws in a very large haystack. That analogy, however, implies that we are indeed seeking but a few straws. In fact, when information arrives on our desks, we often find those few straws are actually far too big and far too numerous for one person to grasp and use easily. In the jargon used in the information world, efficient retrieval of relevant information often results in information overkill.

Ever since writing was invented, it has been a common practice for men to record and store information; not only fact and figures, but also theories and opinions. The rate of recording accelerated after the invention of printing and moveable type, not because that in itself could increase the amount of recording but because, by making it easy to publish multiple copies of a document and sell them at a profit, recording and distributing information became very lucrative and hence attractive to more people. On the other hand, men and women in whose lives the discovery of the handling of information plays a large part usually devise ways of getting what they want from other people rather than from books in their efforts to avoid information overkill. Conferences, briefings, committee meetings are one means of this; personal contacts through the 'invisible college' and members of one's club are another. While such people do read, some of them voraciously, the reading of published literature, including in this category newspapers as well as books and journals and even watching television, may provide little more than 10% of the total information that they use.

Computers have increased the opportunities, not merely by acting as more efficient stores and providers of certain kinds of information than libraries, but also by manipulating the data they contain in order to synthesize new information. To give a simple illustration, a computer which holds data on commodity prices in the various trading capitals of the world, and also data on currency exchange rates, can be programmed to indicate comparative costs in different places in one single currency. Computerized data bases, i.e. stores of bibliographic information, are now well established and quite widely available for anyone to use. Also increasing are the number of data banks, i.e. stores of factual information, which are now generally accessible. Anyone who buys a suitable terminal may be able to arrange to draw information directly from these computer systems for their own purposes; the systems are normally linked to the subscriber by means of the telephone network. Equally, an alternative is now being provided by information supply services such as libraries, more and more of which are introducing terminals as part of their regular services.

The number of sources of information on any topic can therefore be very extensive indeed; publications (in the widest sense), people (experts), specialist organizations from research associations to chambers of commerce, and computer stores. The number of channels by which one can have access to these vast collections of information are also very numerous, ranging from professional literature searchers, via computer intermediaries, to Citizens' Advice Bureaux, information marketing services and information brokers.

The aim of the Guides to Information Sources (formerly Butterworths Guides to Information Sources) is to bring all these sources and channels together in a single convenient form and to present a picture of the international scene as it exists in each of the disciplines we plan to cover. Consideration is also being given to volumes that will cover major interdisciplinary areas of what are now sometimes called 'mission-oriented' fields of knowledge. The first stage of the whole project will give greater emphasis to publications and their exploitation, partly because they are so numerous, and partly because more detail is needed to guide them adequately. But it may be that in due course the balance will change, and certainly the balance in each volume will be that which is appropriate to its subject at the time.

The editor of each volume is a person of high standing, with substantial experience of the discipline and of the sources of information in it. With a team of authors of whom each one is a specialist in one aspect of the field, the total volume provides an integrated and highly expert account of the current sources, of all types, in its subject.

D. J. Foskett
Michael Hill

Contributors

R. T. Adkins MIInfSc
Consultant Information Scientist, Hatfield, Hertfordshire

Dr. R. J. Crawford PhD DSc CEng FIMechE FPRI
Department of Mechanical and Manufacturing Aeronautical, and Chemical Engineering, The Queen's University of Belfast

R. J. E. Cumberbirch BSc
Group Manager, Information Services, British Textile Technology Group, Manchester

S. Catherine Haworth BSc MA ALA
Librarian, Paint Research Association, Teddington, Middlesex

E. Inglis
Chief Translator, RAPRA Technology Ltd, Shawbury, Shropshire

K. P. Jones FLA MIInfSc
Head, Information Group, Tun Abdul Razak Laboratory, The Malaysian Rubber Producers Research Association, Brickendonbury, Hertfordshire

Dr. P. Käfer
Bayer AG, Zentrale Forschung und Entwicklung, Wiss. Information und Dokumentation, Leverkusen

John R. Lawrence
Independent Consultant and former Technical Director, the Society of the Plastics Industry, Wilton, Connecticut

Professor R. G. Linford MA DPhil CChem FRSC
Designate Head of School of Applied Physical Sciences, Leicester Polytechnic

Moyra B. McAllister BSc DipLib
Librarian, Victorian Parliamentary Library, Mt Waverley, Victoria

Dr. I. Omae
Teijin Tokyo Research Centre, Teijin Limited, Tokyo

Professor R. A. Pethrick BSc PhD DSc CChem FRSC
Department of Pure and Applied Chemistry, University of Strathclyde, Glasgow

Dr. L. Schneider
Marketing Division, OMIKK, (National Technical Centre and Library), Budapest

J. A. Shelton MPRI
Information Officer, Fulmer Yarsley Limited, Redhill, Surrey

Dr. J. Sibley
Patents Department, Shell International Petroleum Co Ltd, Shell Centre, London

Judy Stubbington BLib
Former Librarian, RAPRA Technology Ltd, Shawbury, Shropshire

Dr. I. Thomas
Information Services, Royal Society of Chemistry, The University, Nottingham

John S. Widdowson
Project Manager — Information, RESOURCE, c/o British Standards Institution, Milton Keynes

CONTENTS

PART I

CHAPTER ONE

Introduction

R. T. ADKINS

Although their chemical and physical natures have only gradually been revealed during the present century, polymeric materials of organic origin have been known and used for millennia. In the familiar forms, amongst many others, of fibres, waxes, resins, gums, horn and ivory harvested from plants and animals they have proved their value over the years. Indeed, life itself to a considerable degree depends on the ability to form unique polymers and their properties, ranging from massive ligno-cellulosic structures in the plant kingdom to the subtle sequences of aminoacids and sugars on which all life forms are nurtured, perpetuated and modified.

Inorganic polymers however, for example glass and ceramics, while also having a technological base spanning the centuries and a considerable literature in their own right, fall somewhat outside of the scope of this guide and we do not do them full justice, but nevertheless some of the sources discussed in this work may be of interest to those seeking information upon these materials.

Synthetic polymers initially modelled on the natural materials they are tending to replace have been gradually accepted as essential materials with their own unique characteristics. They are no longer inferior substitutes at best for articles where traditional wood, metals, paper, leather etc. (instances may be multiplied a hundred-fold), have been used, but exciting materials opening up new possibilities for exploitation in every walk of life.

Since the possibilities are so numerous the sheer variety of applications brings its own problems. Although soundly science based in its orientation, and as a result one might think, susceptible to the kind of information management techniques which have proved reasonably successful with other disciplines, information on plastics and polymers has grown at a much faster rate than the scientific literature as a whole, which may be the root cause of many of the difficulties apparent in handling the written word in the polymer field. Since the materials are so recent, existing classification schemes for handling the literature have tended to lack precise detail in recognition of new chemical and physical concepts. The Universal Decimal Classification (UDC), for example is somewhat deficient in classes for many now well established polymer chemical types, processes and applications. To be fair, many other systems have also found difficulty in coping with the rapidity of developments, typically tending to lack specificity for the earliest appearance of new concepts and changing the system only when the pressure of growth becomes an irresistible force for change. The number of classification systems which subsume plastics and polymers (a phrase also not without ambiguity), under the blanket heading of 'chemical and allied industries' is still a source of difficulty and needless expenditure of search time! One of the key chapters discusses the problems of nomenclature, providing fresh insights into the different approaches which have been made, whether based on starting materials or the polymer structure itself.

It will be an important aim of the book to try to indicate differences in quality of services and sources, not always easy or in fact entirely desirable, good information often being found in the most unlikely sources. It is probably safe to say that the perfect system for handling information has yet to be devised. Modern information retrieval technology has made tremendous advances in coping with the flood of information, but as with all service industries compromises have to be made. Usually these are reflected in the quality of the indexing of the material, in the interests of providing reasonable value for money while at the same time remaining competitive. Since the plastics and polymer industry, and the information sources on which it depends, have long been international in scope, an attempt has been made to cover all the major producing regions of the world to give some insight into the information resources which may be available uniquely from these areas. The authors of these chapters have had complete freedom to describe the sources available from the regions they cover, and it is hoped that a reasonably comprehensive view of the world's literature sources will be the result.

The book has been structured into three sections: (i) specific source types; (ii) subject sources; (iii) national and regional sources. Although some duplication of coverage may have resulted the different perspectives given by the individual contributors compensate to some degree for this. It is hoped the book will be of value to all those who have a need for information in this wide ranging field, student, reference librarian, information scientist or end-user, at whatever level of expertise they bring to the search.

CHAPTER TWO

Serial literature — primary journals

J. STUBBINGTON

Primary journals may be defined as serial publications appearing at regular intervals of less than a year under the same title (bearing in mind however, the many titular changes which occur through merger or separation), containing new knowledge or new interpretations of existing knowledge. The word knowledge itself may cause semantic difficulties but for the purposes of this chapter is not limited to novel technical knowledge but includes a wide range of other material whose potential usefulness depends in equal measure on its organization, as in for example an alphabetical listing of company names under a product heading, as on its novelty.

No less than in other fields the form and characteristics of the primary literature in polymers and plastics is on the whole well adapted to the needs of the users of the literature in terms of speed of publication, packaging the information in discrete portions permitting practicable measures to be taken for its dissemination, storage and retrieval. Although experiments are in progress with publication of research material in electronic form (*Chemical Journals Online*, the major primary journals published by the Royal Society of Chemistry) it is unlikely that the printed version will ever be totally supplanted by the newer media. Periodicals specialising in the polymer and plastics field have tended to keep pace, if not actually outstrip the growth in scientific literature generally, particularly since the second world war.

As a generalisation two fairly distinct categories of journal have emerged over the years. First, there is the traditional academic journal focusing on new research, often still published by the institutions continuing to support such work but increasingly by international specialist publishers able to satisfy the demand for publication by the scientific community. These journals frequently have a long history and are well established and respected internationally. As well as research papers, reviews and editorials are a prominent feature. The second category consists mainly of journals published largely for commercial reasons and targetted mainly on industrial research, development and business aspects of the industry for the people who are involved in these functions. The journals carry, in addition to much information on process development work, book reviews, calendars of forthcoming meetings and events, new trade literature and a considerable amount of advertising matter. Some of the titles generally considered to be important are discussed below but of necessity the choice will be subjective and limited. More specialised titles are considered in their relevant chapters relating to a specific aspect of the industry.

Angewandte Makromolekulare Chemie (Hüthig and Wepf Verlag). This title appeared as an offshoot of the well established *Makromolekulare Chemie* in 1969, publishing fifteen to twenty articles monthly on the (physical) chemistry of polymer production, fabrication processes and applications. Coverage is worldwide and articles may appear in German, French or English.

British Polymer Journal (Society of the Chemical Industry). This title has been produced monthly since 1969 by the Macro Group, as a joint venture of the Royal Society of Chemistry and the Society of the Chemical Industry. Academic and industrial aspects of biopolymers, polymer chemistry and physics and applied polymer science are published, mainly but not exclusively by British authors. Book reviews and a calendar of events also appear.

Colloid and Polymer Science (Dr Dietrich Steinkopff Verlag). Subtitled the *Official Journal of the Kolloid Gesellschaft* and published in English the journal is produced in two sections: *Colloids* and *Polymer Science*. Each monthly issue focuses on significant developments in polymer synthesis, polymerization processes and special polymers. Book reviews appear in most issues.

European Polymer Journal (Pergamon Journals Ltd.). Published monthly since 1965 this journal carries a wide range of articles in a number of European languages.

International Journal of Polymeric Materials (Gordon and Breach, Science Publishers Inc.). This was launched in 1971 as a monthly international journal covering plastics, fibres, composites and elastomers with the intention of promoting reliable information on polymer properties for chemists, physicists, engineers and designers.

Articles are in English with the emphasis on polymers useful as engineering materials, their property dependence on structure, morphology, processing and their suitability for particular end-users and environments.

Journal of Applied Polymer Science (John Wiley and Sons Inc.). This journal has appeared since 1959 and is published in two sections, the major part as sixteen issues per year while the sub-section *Applied Polymer Symposia* is issued irregularly. Articles of high international repute are published on polymer analysis, characterization, techniques and instrumentation relating to natural and synthetic polymers.

Journal of Elastomers and Plastics (Technomic Publishing Co. Inc.). This is a quarterly journal covering activities in the USA in depth, concentrating on developments in elastomer technology such as blending, block copolymerization, filled and reinforced elastomers. A regular patents digest is also included.

Journal of Macromolecular Science (Marcel Dekker Inc.). This journal was established in 1967 and is published in three parts:

A — Chemistry, (monthly) usually contains about eight papers on polymer chemistry and occasional book reviews.

B — Physics, is published four times a year, each issue containing papers on the fundamentals of polymer physics in the solid and liquid state contributed by international authors.

C — Reviews in Macromolecular Chemistry and Physics, is also published four times a year and contains authoritative reviews, particularly of recent developments.

Journal of Materials Science (Chapman and Hall). Although broader in scope, as its title implies, than the specialist polymer journals, the journal and its cognate publication *Journal of Materials Science Letters* containing short items of usually less than 1500 words, are important sources in this subject area. Publication started in 1969 and monthly issues contain papers and reviews on the relationship of structure, properties and end-uses of materials.

Journal of Polymer Science (John Wiley and Sons Inc.). Originally published as a single title in 1946, sub-division into sections reflecting both the increasing tendency towards specialisation and the rapid growth rate in polymer research has resulted in the current four-part series.

Polymer Chemistry: About forty papers on the chemistry and physical chemistry of polymers are published monthly and a smaller number of notes, shorter papers of narrower scope.

Polymer Letters: Eight brief articles are published monthly for the purposes of rapid communication of preliminary and intermediate experimental results, and of establishing priority in the field under investigation.

Polymer Physics: This is the monthly counterpart to *Polymer Chemistry*, containing about fifteen articles and notes on high polymer physics.
Polymer Symposia: This appears irregularly, being dependent on the marshalling of conference proceedings, a usually protracted activity but important in capturing material which might not otherwise be disseminated widely.
Macromolecules (American Chemical Society). This is another journal having its origins in the late sixties, a period which saw tremendous growth in research and development in polymers and plastics. About twenty articles, notes, communications to the editor and one major review are published monthly on synthesis, polymerization mechanisms and kinetics, chemical reactions, and properties of organic, inorganic and biopolymers in solution and in bulk.
Makromolekulare Chemie (Hüthig and Wepf Verlag). Some twenty to twenty-five articles in German, English and French are published monthly, distributed between the three parts, *Makromolekulare Chemie: Rapid Communications* and *Macromolecular Symposia*. In many respects its history mirrors its USA counterpart *Polymer Chemistry*, it first appeared in 1947 and from the outset the coverage of theoretical and experimental chemistry, physical chemistry and physics was international in scope.
Materiale Plastice Inst. Cent. Cercetari Chim. This is a Romanian monthly, occasionally irregular publication, containing about ten articles in each issue. Perusal will give some insight into developments and trends in a part of Eastern Europe rich in resources and of increasing commercial importance. The language is an impenetrable barrier to the majority, but English abstracts appear in *RAPRA Abstracts* and selectively in *Chemical Abstracts*. Translation problems with minority languages are discussed in a later chapter.
Polymer (The International Journal for the Science and Technology of Polymers) (Butterworth Scientific Ltd.). Established in 1960 this monthly journal publishes about thirty papers each month, book reviews, a calendar and letters. A supplement *Polymer Communications* also appears monthly, carrying brief reports of current work and conclusions arising from completed work, for rapid dissemination.
Polymer Bulletin (Springer Verlag). Appearing monthly since 1978, this journal is published in English and is international in scope with the aim of rapid publication of significant advances in materials science, including the chemistry, physical chemistry and physics of synthetic and biopolymers.
Polymer Engineering and Science (Society of Plastics Engineers Inc.). In 1965 this journal was launched as a semi-monthly (except

for monthly issues in June and December) continuation of *SPE Transactions* which started in 1961. International in scope, although the tendency to favour American and Japanese authors is perhaps merely a reflection of the vast output of research from these two countries demanding publication.

Polymer Journal (Society of Polymer Science, Japan). A monthly journal started in 1979 it publishes a selection of development work deemed significant by The Japanese Society of Polymer Science. Abstracts of the society's other journal, *Kobunshi Ronbunshu* are included.

Polymer-Plastics Technology and Engineering (Marcel Dekker Inc.). Originally *The Journal of Macromolecular Science, Reviews in Polymer Technology, Part D*, it is published quarterly as a vehicle for critical reviews in plastics technology and engineering.

Polymer Process Engineering (Marcel Dekker Inc.). This monthly journal covers theoretical as well as experimental aspects of polymer technology and engineering.

Polymer Science USSR (Pergamon Press Ltd.). This journal is a valuable cover-to-cover translation into English of *Vysokomolekulyarnye Soyedineniya Series A*. Published monthly some considerable time after the original the journal provides useful access to developments within the USSR.

Progress in Colloid and Polymer Science (Dr Dietrich Steinkopff Verlag). Appearing somewhat irregularly in hard-backed covers this publication reviews aspects of polymer science under the guidance of invited specialist editors.

Rubber Chemistry and Technology (American Chemical Society, Rubber Division). One of the older journals, established in 1928, it has a well established reputation for coverage of theoretical, and practical aspects of the subject. It is issued bimonthly, excepting for January and February, and one of the five annual issues contains a valuable summary of current developments and a selection of Rubber Reviews for the year. A meetings and events calendar, letters, and book reviews are also included.

Specialist subject periodicals

Other more specialised journals are discussed briefly below. On the whole, they follow the pattern of the more general publications, with the difference that they aim for a narrower, specialist market, and are thus virtually required reading for the practitioners in these areas.

Composites (The International Journal of the Science and Technology of Reinforced Materials) (Butterworth Scientific Ltd.). This monthly journal covers research and development of new materials and applications in plastics, cement, metals and ceramics. A useful literature survey and patents digest is also included.

Journal of Biomaterials Research (John Wiley and Sons Inc.). Containing papers on original work and reviews, each quarterly issue concentrates on a specific clinical area covering industrial, academic and governmental research aspects, and also contains feature articles on new health care opportunities.

Journal of Cellular Plastics (Technomic Publishing Co. Inc.). Issued bimonthly, this title covers the chemistry, formulation, processing, properties and performance of expanded plastics. A patents digest and an industry section carrying news items on product development, chemical components, additives, standards, new machinery, company news, meetings, market analyses and forecasts is also included.

Journal of Coated Fabrics (Technomic Publishing Co. Inc.). This quarterly journal covers studies on fabric technology, processes, products and performance. Special areas of interest include fire behaviour and retardance, testing, standards, markets and new applications. A patents digest is also available.

Journal of Composite Materials (Technomic Publishing Co. Inc.). This bimonthly journal is a valuable source of new data on materials design, analysis, testing, performance and new applications.

Journal of Fire Sciences (Technomic Publishing Co. Inc.). Published bimonthly, the journal carries in-depth technical studies of flammability, chemistry of combustion, toxicology, fire scenarios, fire retardance, test methods, standards and requisites.

Journal of Plastic Film and Sheeting (Technomic Publishing Co. Inc.). A useful quarterly journal covering co-extrusion, barrier properties, shrink and stretch packaging techniques, vacuum forming, aseptic packaging and film modifications. High performance films are also covered.

Journal of Reinforced Plastics and Composites (Technomic Publishing Co. Inc.). Appearing quarterly, this journal presents research papers on composition, design, reinforcement, analysis properties and performance with considerable emphasis on environmental effects on composites and on their non-destructive testing and failure patterns.

Journal of Thermal Insulation (Technomic Publishing Co. Inc.). This quarterly journal is concerned with new developments in materials, properties, performance, test methods, equipment and applications. Standards, regulations, new legislation and a calendar of meetings are also included.

Polymer Composites (Society of Plastics Engineers). Published bimonthly this journal is similar in scope to *The Journal of*

Reinforced Plastics and Composites.

Reactive Polymers (Elsevier Scientific Publishing Co.). Published monthly the journal covers novel and unusual materials, fluid compositions, catalysts, synthesis, methods and applications with data on theoretical, design and processing aspects.

Reinforced Plastics (McDonald Publications of London Ltd.). This monthly journal is particularly useful for book reviews and a buyers guide to reinforced plastic materials and moulders sundries.

Journals whose primary function is not the publication of original research work, but whose importance perhaps, lies mostly in supporting awareness of industrial achievement and market developments, form an essential part of the literature. It is, after all, necessary to know what competitors in the industry are doing, a need met by the wide circulation this category of popular and important journals receives. Emphasis is largely on reporting process and product developments, and providing a medium for discussion of events and activities of individuals, companies and governments which have an impact on trade. The titles discussed below have been selected as examples of this category and is not intended to be exhaustive.

British Plastics and Rubber (McMillan Publishing Ltd.). This monthly journal incorporates the previously separate publications *Polymer Age* and *Rubber and Plastics Age* and is notable for its high standard of feature articles containing technical data. An annual directory of suppliers and services is also included.

Chemical and Engineering News (American Chemical Society). This weekly publication frequently contains items on new developments in polymers and plastics.

Chemical Week (McGraw-Hill Inc.). Brief reports are carried on new research, markets and economic trends, in addition to a major feature article.

Elastomerics (Comunication Channels Inc.). Formerly *Rubber Age*, there are now two editions, one for the USA market and an overseas edition.

European Chemical News (IPC Industrial Press Ltd.). This weekly journal gives extensive coverage to market developments in the plastics field, highlighting specific areas with special reports, and including details on new research projects.

European Plastics News (Business Press International Ltd.). Incorporating *British Plastics* and *Europlastics* this monthly journal features detailed reports on materials, machinery and applications developments.

European Rubber Journal (Crain Communications Ltd.). This monthly contains some useful 'special reports' and product profiles.

Japan Chemical Week (The Chemical Daily Co. Ltd., International Division). Despite its obvious concern to concentrate on developments in the Japanese industry this monthly journal is of great interest to the wider, international audience, carrying features on materials applications and machinery, and including an annual directory.

Modern Plastics International (Modern Plastics International). This popular monthly has earned its high reputation for the relevant and detailed special reports accompanying feature articles on product and process engineering, manufacture, research and development. New patents are also included.

Plastics and Rubber International (Plastics and Rubber Institute). This bimonthly is the official journal of the PRI and a useful source for new polymer developments, fabrication processes and machinery. An annual buyers guide is also produced.

Plastics and Rubber Weekly (Maclaren Publishers Ltd.). This well known and highly regarded weekly newspaper gives wide ranging topical coverage of commercial and technological developments, reports on new equipment, standards and quality control. Meetings, exhibitions and trade fairs are advertised and reported, and new books reviewed.

Plastics Engineering (Society of Plastics Engineers Inc.). Published monthly this is a valuable source of SPE news, covering new research and development project details, processing, additives, design, finishing and applications. Patents, trade literature and forthcoming meetings are advertised.

Plastics Industry News (Japan) (Institute of Polymer Industry). Monthly production figures are a useful feature of this journal, together with information on new products and industrial developments, a calendar of forthcoming conferences and exhibitions and book reviews.

Plastics Technology — the magazine of plastics manufacturing productivity (Bill Communications Inc.). This is a monthly journal containing price updates, together with the usual news items and features and an annual handbook and buyers guide are also included.

Plastics World (Cahners Publishing Co. Inc.). For the manager in the plastics industry, this monthly journal carries information on new materials, applications and machinery, supplemented by business news. An extra issue is published in March, and an annual directory is also produced.

Rubber and Plastics News (Crain Automotive Group Inc.). Published twice monthly in newspaper format it covers a broad spectrum of technological and commercial activities, announcements of forthcoming events and new trade literature. This journal is supplemented by *Rubber and Plastics News 2*, featuring brief technical reviews as well as news items.

Rubber Developments (Malaysian Rubber Producers Research Association). Published quarterly this journal describes current developments in natural rubber research, production technology and uses. Book reviews and general MRPRA news are also included.

Rubber World (Lippincott and Peto Inc.). Subtitled as the *Technical Service* magazine for the rubber industry, and appearing monthly, coverage includes new applications, machinery and products, a meetings calendar and patents. The October issue gives details of chemicals and rubber grades. The *Rubber Blue Book* is produced as an annual directory.

Conclusions

The titles mentioned in this chapter represent only a fraction of the primary source journals of potential interest to the user. Increasing specialisation of coverage however, reflects the growth of the industry, and incidentally helps the user to limit the journals which must be scanned in order to stay abreast of developments. Although everyone has their favourites, the choice of journals is nevertheless a matter for frequent reassessment.

To help with this a number of organizations publish useful information on new and current titles from time to time, for example:

British Library Document Supply Centre:
Current Serials Received, obtainable from the Document Supply Centre, Boston Spa, Wetherby, West Yorkshire LS23 7BQ.

CAS Source Index (CASSI) and *CAS Source Index Supplement to CASSI* is a cumulative listing of all periodicals abstracted by the Chemical Abstracts Service, American Chemical Society, Washington, USA.

Deutsches Kunststoffe Institut, *Verzeichnis der im Literatur-Schnelldienst 'Kunststoffe. Kautschuk, Fasern'* DKI, Darmstadt, West Germany.

Plastics and Rubber Institute Publications Catalogue contains a fully indexed list of periodicals, monographs, reviews and conference proceedings. A brief appraisal of PRI publishing policy is given in *Plastics and Rubber International*, **10, 2,** April 1985, 46.

RAPRA Abstracts Annual List of Journals appears in the first January issue of *RAPRA Abstracts.*

RAPRA Publications: Books, Trade Studies, Journals, Translations, Software is published irregularly several times each year and contains full details of the latest publications available to both members and non-members: RAPRA Technology Ltd., Shawbury, Shrewsbury, Shropshire.

Rubber Research Institute of Malaya Library, *List of Journal Holdings*: RRIM, Kuala Lumpur, Malaya.

Salmon, C. *Repertoire des périodiques relatif aux matières plastiques et aux caoutchouc synthétique* (Bibliographia Belgica, Commission Belge de Bibliographie, Brussels).

CHAPTER THREE

Serial literature — abstracts

I. THOMAS

Because of the sheer volume of scientific and technological information published, most scientists and engineers rely on abstracting publications and services to alert them to developments in areas of interest. The use of abstracting publications allows a rapid assimilation of the information content of a far wider variety of sources of information than any working scientist could hope to examine. In addition, abstract publications allow ready access to material in foreign languages. However, perhaps the greatest service performed by abstracting services is that they create indexes of the information that is available in primary and other sources: these indexes are the keys to finding information on any subject.

The largest and most widely accessible abstracting service in the field of polymers and plastics is *Chemical Abstracts* which, because of its size, is also a very difficult publication to use effectively. Much of this chapter is therefore devoted to *Chemical Abstracts* and to ways of finding any desired information within it.

Because nowadays the area of abstracting services is inextricably intertwined with the area of computer data bases (computer data bases are used to produce many abstracting publications and vice versa), a brief section of the chapter deals with computer data bases and data banks relating to polymers and plastics.

Chemical Abstracts

Chemical Abstracts (CA), published by Chemical Abstracts Service (CAS), at Columbus, Ohio, USA, aims to cover the whole of the world's chemical literature and, as such, represents one of the most comprehensive sources of information on polymers and plastics. CA is published weekly and comprises 80 sections, each relating to an area of chemistry: Sections 1–34 appear in odd-numbered issues, Sections 35–80 in even. The majority of polymer-related information appears in sections 35–46:

35. Chemistry of Synthetic High Polymers
36. Physical Properties of Synthetic High Polymers
37. Plastics Manufacture and Processing
38. Plastics Fabrication and Uses
39. Synthetic Elastomers and Natural Rubber
40. Textiles
41. Dyes, Organic Pigments, Fluorescent Brighteners, and Photographic Sensitizers
42. Coatings, Inks, and Related Products
43. Cellulose, Lignin, Paper, and Other Wood Products
44. Industrial Carbohydrates
45. Industrial Organic Chemicals, Leather, Fats, and Waxes
46. Surface-active Agents and Detergents

A few specialised polymer topics do appear outside these sections, e.g. photopolymerizable compositions for printing-plate manufacture appear in Section 74, Radiation Chemistry, Photochemistry, and Photographic and Other Reprographic Processes.

Coverage

CAS examines about 12,000 scientific and technical journals from around 100 countries, together with patent documents, conference and symposium proceedings, monographs, government reports, and scientific books from around the world. Any items falling within a very broad definition of chemistry and chemical engineering are selected for inclusion in CA.

CA abstracts

The CA abstract (Fig. 3.1) is a concise representation of the new information appearing in an item in a primary journal or other source. The heading identifies the document and includes the title, author(s) name(s), work location of the first author cited, bibliographic citations, and the language of the original document. The body of the abstract summarises the new findings contained in the original document and aims to give the reader sufficient information to decide if it is worthwhile obtaining the original document: it is not intended to replace that document, nor does it represent any evaluation of the scientific and technical merits of the original item.

104: **6236u Chiral complexes polymerize methacrylate esters to give helical polymers that mutarotate by uncoiling.** Cram, Donald J.; Sogah, Dotsevi Y. (Dep. Chem. Biochem., Univ. California, Los Angeles, CA 90024 USA). *J. Am. Chem. Soc.* **1985,** 107(26), 830–2 (Eng). Three enantiomerically pure chiral crown ether hosts based on the 2,2'–disubstituted 1,1'–binaphthyl unit form complexes with various organometallic guests in **PhMe** at −78°. The complexes initiate anionic polymn. of Me, *tert*–Bu, or benzyl methacrylate to produce 80–90% isotactic polymer of no.–av. mol. wt. 480–2100 with optical rotations of $[\alpha]_{578}^{25}$ (THF, c2) that range from −250° to +350°. The polymers from Me and *tert*–Bu methacrylate mutarotate within hours at ambient temps. to rotations close to zero. The polymer from benzyl methacrylate was optically stable over the same time span. The high rotations are attributed to helicity, and the mutarotation to randomization of the conformations of the polymers. Models for the propagation step are suggested which correlate the facts and predict the configurations of both the chiral centers and the helices.

Figure 3.1

Because CA contains almost 500,000 abstracts a year, a number of access routes are available to enable a user to identify those items relevant to a particular topic. Some of these access routes appear in the weekly issues of CA; others appear in semiannual indexes.

CA Indexes

KEYWORD INDEXES

Keyword Indexes appear at the back of each weekly issue of CA and can be used to locate items on a given topic in that particular issue. Each item in CA will, in general, have several accompanying entries in the Keyword Index. Keyword Indexes are natural language indexes that use both author terminology and the terminology current in the field. Since the vocabulary is not standardized, a variety of synonyms is usually required in searching. For example, a searcher interested in poly(tetrafluoroethylene) should also look in the Keyword Index at such terms as PTFE or Teflon.

Keyword Indexes are not in-depth indexes: they do not necessarily refer to every subject or substance contained in the original document. However, each semiannual volume of CA is accompanied by three in-depth indexes: the Chemical Substance Index, the Formula Index, and the General Subject Index.

CHEMICAL SUBSTANCE INDEX

The Chemical Substance Index provides comprehensive access to every specific chemical substance for which information is presented in the original item: this can mean that there are, in some cases, hundreds of entries for a particular item. Each substance in the Chemical Substance Index is assigned a systematic name, the CA Index Name, so that all the items relating to a particular substance appear in one place in the index. The substances in the Chemical Substance Indexes form the CAS Registry File; this is a computer file of all the unique substances indexed by *Chemical Abstracts*. Each unique substance is assigned a serial number — the CAS Registry Number — which may then be used as a concise way of representing any chemical substance. The Registry Number has no physical significance; it is, in effect, a catalogue number. Each unique material has its own Registry Number. This means that all unique polymers, including polymers of differing stereochemistry (tacticity) or isotopically labelled polymers, have their own Registry Numbers.

Polymer nomenclature in CA, in general, is based on that of the constituent monomer(s). In the case of a homopolymer, the CA Index Name of the monomer is followed by the modifying term 'homopolymer', e.g. polystyrene appears as 'benzene, ethenyl-, homopolymer [9003–53–6]' (9003–53–6 is the CAS Registry Number). For copolymers, the CA Index Name is followed by the phrase 'polymer with . . .' and the names of the other monomers; e.g. butadiene-styrene copolymer appears at 'benzene, ethenyl-, polymer with 1,3-butadiene [9003–55–8]' *AND* at '1,3-butadiene, polymer with ethenylbenzene [9003–55–8]'. This multiple indexing of copolymers allows the user to find the homopolymer(s) and all copolymers of a particular monomer at the same alphabetical location in the index.

Oligomers. It must be stressed that *all* polymers from a particular monomer, or set of comonomers, *regardless of molecular weight*, appear at the same heading in the Chemical Substance Index: the only exception is that of oligomeric compounds of a known structure and/or degree of polymerisation, which will appear at the systematic name of the compound (e.g. the cyclic dimer of 1,3-butadiene appears at 'cyclohexene, 4-ethenyl- [100–40–3]') or at the name of the monomer with an appropriate modifying phrase (e.g. a 1-hexene dimer of unknown structure appears at '1-hexene, dimer [18923–86–9]'). Oligomers of unknown (or variable) degree of polymerisation and structure appear under the heading for the corresponding polymer with the modifying phrase 'oligomeric'.

Block, Graft, and Alternating Copolymers. Up to January 1987, CAS indexed all copolymers in an identical fashion: e.g. all the copolymers of 1,3-butadiene and styrene appeared in the same place in the indexes and were assigned the same Registry Number [9003–55–8], regardless of their nature. Details (block, graft, etc.) were given in subsidiary phrases. From 1 January 1987, however, CAS decided to index block, graft, and alternating copolymers differently and to assign unique Registry Numbers to each. The original Registry Number will continue to apply to the random copolymer. Users need to be aware of this change, especially when conducting a retrospective search.

Post-treated Polymers. Post-treated polymers are those modified after polymerisation has occurred. Post-treated polymers are indexed as unmodified polymers with the modification indicated in an auxiliary phrase. Depending on the type of modification, they may be treated as unique substances and accorded unique Registry Numbers, or may be treated as incompletely defined derivatives of the parent polymer.

Structural Repeating Units. Since it is possible to make the same polymer from different monomers (see Fig. 3.2) or to make several polymers from one monomer or set of monomers (Fig. 3.3), it is useful to have access to polymer information on the basis of structures of polymers: this access takes the form of the Structural Repeating Unit (SRU). Where the structure of a polymer has been unambiguously characterized (with supporting data), or where a single structure can safely be inferred from the chemistry, CAS assign an SRU to the polymer; this may supplement or, in some unusual cases, replace the monomer-based index entries. SRU names are not provided routinely for addition polymers. However, where an author provides data to substantiate the structure of an additional polymer, the SRU will be assigned; for example, if an author prepared polyacrylamide and substantiated its structure, an additional index entry would appear at the heading 'poly[imino(1-oxo-1,3-propanediyl)] [24937–14–2]'. SRU names are not assigned to polymers prepared from unsymmetrical monomers or condensation polymers of 3 or more monomers.

$H_2N(CH_2)_6NH_2 + HO_2C(CH_2)_6CO_2H$

$H_2N(CH_2)_6NH_2 + ClCO(CH_2)_6COCl \longrightarrow [NH(CH_2)_6NHCO(CH_2)_6CO]_n$

Nylon 68

$H_sN(CH_2)_6NH_2 + MeO_2C(CH_2)_6CO_2Me$

Figure 3.2

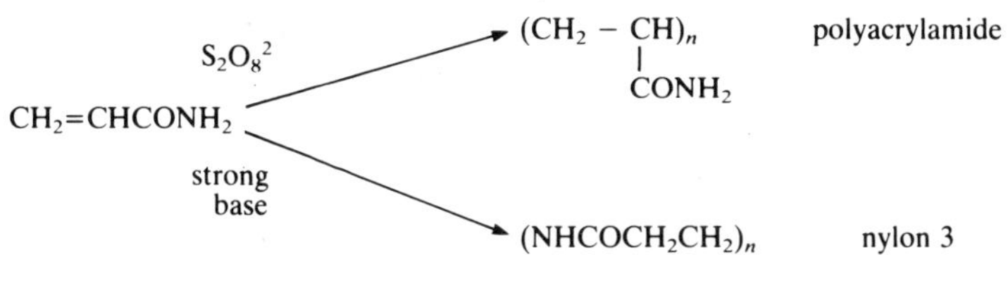

Figure 3.3

There are several very common polymers where the SRU index entry replaces monomer-based entries. These are:

Polymer	*Indexed at SRU Name*
1. Polycaprolactam 6-Aminohexanoic acid homopolymer Nylon 6 ε-Caprolactam polymer	Poly[imino(1-oxo-1,6-hexanediyl)] [25038–54–4]
2. Nylon 66 Adipic acid-1,6-hexanediamine polymer	Poly[imino(1,6-dioxo-1,6-hexanediyl)imino-1,6-hexanediyl] [32131–17–2]
3. Polyethylene glycol Polyoxyethylene Polyethylene oxide	Poly(oxy-1,2-ethanediyl), α-hydro-ω-hydroxy- [25322–68–3]
4. Polypropylene glycol Polyoxypropylene Polypropylene oxide	Poly[oxy(methyl-1,2-ethanediyl], α-hydro-ω-hydroxy- [25322–69–4]
5. Poly(ethylene terephthalate) PET	Poly(oxy-1,2-ethanediyloxy-carbonyl-1,4-phenylenecarbonyl) [25038–59–9]

Note that for polyethylene glycol and polypropylene glycol, the end groups are given in the SRU; SRU names for ethers and other derivatives are modified appropriately: e.g. polyethylene glycol dimethyl ether appears at 'poly(oxy-1,2-ethanediyl), α-methyl-ω-methoxy- [24991–55–7]'.

SRU names begin with the term 'poly' followed by the systematic names of the multivalent radicals which make up the structure, in a preferred order derived from a defined set of rules.

Because CA names are often very complex to deduce, most users access the Chemical Substance Index through the CA Formula Index.

THE CA FORMULA INDEX

The CA Formula Index links molecular formulae (in the Hill System order; C, H, then other elements alphabetically) of chemical substances indexed in the Chemical Substance Index to the corresponding CA abstracts. For polymers, the entry points are the molecular formulae for the corresponding monomers: this provides a ready access to a monomer and all its polymers at a single location in the formula index. SRU Formula Index entries are found at the molecular formula entry for the repeating unit, enclosed in parentheses and followed by an italic subscript '*n*'. Since molecular formulae are not unique, the user must decide which name found under a particular molecular formula refers to a particular substance.

THE GENERAL SUBJECT INDEX

Access to more general information in CA is provided by the General Subject Index. The index is arranged according to a set of controlled terms which appear in a bold typeface; each term is followed by free-text information giving the context in which that particular subject appeared in the original paper. Like the Chemical Substance Index, the General Subject Index is an in-depth index of all the information that appeared in the original document.

Because of the size and complexity of CA, there are a number of aids to help a user locate particular information.

THE CA INDEX GUIDE

The CA Index Guide is an essential tool for any user wishing to make the best of a search in *Chemical Abstracts*. The body of the Guide consists of a series of notes and cross-references indicating how both substances and general subjects are indexed in CA, and giving the preferred index terms and useful alternatives to consult (Fig. 3.4). The Guide is updated and reissued every 18 months. In addition to the list of indexing notes and cross-references, the Index Guide contains a hierarchical list of general subject headings, a guide to the organisation of and use of CA Indexes, notes on the selection of General Subject Headings, and a concise guide to CA chemical nomenclature.

It is important when conducting a retrospective search through CA that the appropriate Index Guide for a particular set of volumes is consulted, as indexing policies have changed over the years.

COLLECTIVE INDEXES

Since its inception in 1907, CA has cumulated its indexes regularly to give, initially, 10-year and, since 1957, 5-year Collective

Polyester fibers
Fibers from terephthalic acid polyesters with glycols alone or including other components and from unspecified polyesters are indexed at this heading. Fibers from specified polyesters from acids other than terephthalic acid are indexed at *Synthetic fibers*
Polyester fibers, uses and miscellaneous
dyeing of — see *Dyeing*
Polyesterification
See *Polymerization*
Polyesters
Entries are made at this heading only for ester group containing polymers formed by the polycondensation of polycarboxylic acids, acid halides, or anhydrides with polyhydric alcohols or phenols, or by polymerization of lactones or hydroxy acids. Polymers formed from unsaturated estern, such as butyl 2-propenoate or ethenyl propanoate, by addition polymerization are indexed at the named esters or at *Esters, polymers*
See also *Alcohols, compounds*, reaction products
alkyd — see *Alkyd resins*
carbonate — see *Polycarbonates*
fibers
see
Polyester fibers
Synthetic fibers
polyethylene terephthalate — see *Poly(oxy-1,2-ethanediyloxycarbonyl-1,4-e phenylenecarbonyl; [25038-59-9]*
polyurethane — see *Urethane polymers*, polyester
Polyesters, uses and miscellaneous
coatings, for decoration, finishing or protection — see *Coating materials*
coating with, for decoration, finishing or protection — see *Coating process*

Figure 3.4

Indexes. This facilitates searching of CA, in that material for a 5-year (or 10-year) period can be consulted in one, albeit massive, set of volumes. The CA Collective Index is claimed to be the largest regularly produced publication: the 11th Collective Index, published in 1987, comprises some 93 books with about 162,000 pages.

The 5-year Collective Index span is important to searchers in CA, because major changes in indexing and other policies are generally made at the start of a Collective Index period. The most noteworthy of these changes was a radical change in chemical nomenclature which occurred at the start of the ninth collective period in 1972. Literature sources which cite systematic names will often give both the name used in the 9th and subsequent Collective

Indexes, and the name used in 8th and earlier Collective Indexes, e.g. 2-propenoic acid, 2-methyl-, methyl ester (9CI), methacrylic acid, methyl ester (8CI).

CA SELECTS

To save users from scanning entire sections of CA, CAS produce fortnightly bulletins on very specific topics. The material for the bulletins is selected from the main body of CA by computer search routines. The CA Selects bulletins cover over 164 topics, 28 of which relate to polymer chemistry. The topics include: polyacrylates; fibre-reinforced plastics; polymerization kinetics and process control; and water-based coatings. CA Selects are a very economical way of maintaining awareness of new developments in a particular field.

REGISTRY HANDBOOK — COMMON NAMES

This is a microform publication (Fig. 3.5) which enables a user to identify substances and their Registry Numbers from a variety of common names. The Names Section (left) lists over 531,000 common names for over 335,000 substances and links the names to the corresponding Registry Numbers. The Registry Numbers are linked to the CA name, and other synonyms by the Number Section (right).

CHEMICAL ABSTRACTS SERVICE SOURCE INDEX (CASSI)

Having located a particular item in CA, the user may want to obtain the original document. CASSI contains complete identifying information on more than 60,000 serial and nonserial publications and indicates which of 399 libraries in 28 countries hold these publications. The publication is updated by quarterly supplements. The listings include all the titles covered by CA since 1907, plus additional titles covered by *Chemisches Zentralblatt* 1830–1940 and Beilsteins *Handbuch der Organischen Chemie* to 1965. In addition, CAS offer a document-delivery service, providing photocopies of most documents abstracted by CAS.

ONLINE ACCESS

Most of the information in the CAS data base is available in computer-readable form. The principle files include: CA SEARCH, which contains complete bibliographic information and

Kentanium K162B [12604-75-0]
Kentrol [11138-66-2]
Kentrolite [12281-09-3]$H_{10}O_9Si_2$ 2Mn-2Pb
Kenyaite [12285-95-9]$H_{10}O_{49}Si_{22}$-6H_2O-2Na
Kenyon-Banfield radical [3315-40-0] $C_{18}H_{21}N_2O_2$
KEP 1 [52628-02-1]
KEP 2 [60842-58-2]
KEP 3(B) [60842-59-3]
Kephalin [1405-71-6]
Kephrine [99-45-6] $C_9H_{11}NO_3$
Kephrine hydrochloride [62-13-5] $C_9H_{11}NO_3$-ClH
Kephlon [84-80-0] $C_{31}H_{46}O_2$
KEP 3(M) [60842-60-6]
Kepolen E 163 [9010-79-1] $(C_3H_6\text{-}C_2H_4)x$
Kepolen W [9010-79-1] $(C_3H_6\text{-}C_2H_4)x$
Kepone [143-50-0] $C_{10}Cl_{10}O$
Kepone alcohol [19324-05-1] $C_{10}H_2Cl_{10}O$
Kepone dimethyl ketal [52171-95-6] $C_{12}H_6Cl_{10}O_2$
Kepone hydrate [4715-22-4] $C_{10}H_2Cl_{10}O_2$
Keptan [10405-02-4] $C_{25}H_{30}NO_3$-Cl
Keracyanin [18719-76-1] $C_{27}H_{31}O_{15}$-Cl
Keralec [8061-78-7]
Keralba [498-78-2] $C_{24}H_{42}N_2O_3S$

143-48-6 $C_2H_6O_3Si$
Silanetriol, ethenyl- 9Cl
Silanetriol, vinyl- 8Cl
Etheneorthosiliconic acid
Ethyleneorthosiliconic
Ethyleneorthosiliconic acid
Vinylsilanetriol
Vinyltrihydroxysilane
143-50-0 $C_{10}Cl_{10}O$
1,3,4-Metheno-2H-cyclobuta[cd] pentalen-2-one, 1,1a,3,3a,4, 5,5,5a,5b,6-decachlorooc (ahydro-8Cl,9Cl
Chlordecone
Clordecone
Compound 1189
Decachloroketone
ENT-16391
GC 1189
Kepone
Merex
143-52-2 $C_{18}H_{21}NO_3$
Morphinan-6-one,4,5-epoxy-3-hydroxy-5,17-dimethyl-,(5a)-9Cl
Morphinan-6-one,4,5a-epoxy-3-hydroxy-5,17-dimethyl—cl
Methyldihydromorphinone
5-Methyldihydromorphinone
Metopon

Figure 3.5

the keyword and controlled-vocabulary indexes for all the documents abstracted in CA; Registry Structure Service, which contains the complete structural records for more than 8 million compounds indexed by CAS; and the Registry Nomenclature Service, which contains more than 10 million common, trade, and systematic names on file for substances in the CA registry.

These files have been mounted on several public-access computer systems. A user at a suitably equipped remote computer terminal (or microcomputer) can communicate with the system by way of the telephone system and conduct searches, interactively, on the computer. These systems are becoming increasingly sophisticated, allowing search by chemical structures and substructures, in addition to the more usual textual terms. For a guide to online-searching for polymer information on the CAS ONLINE system, operated by CAS on the STN Network, see *POLYMERS*

ONLINE Searching Polymers by Structures through CAS ONLINE, CAS (1983) and *ACTION GUIDE for Search for Polymers by Structure*, CAS (1983).

Other Abstracting Services in Polymers and Plastics

RAPRA Abstracts

RAPRA Abstracts published monthly by RAPRA Techology Ltd., Shawbury, Shropshire, UK, since 1968 covers information pertinent to the rubber and plastics industries. It is more commercial and technical in orientation than CA, in that it covers such areas as legislation and business information, which CA for the most part omits.

Each issue of *RAPRA Abstracts* contains some 2000 items selected from 400 or so periodicals, conference proceedings, books, technical trade literature, standards, patents, and government publications. The material is presented in 13 sections:

General
Commercial and Economic
Legislation
Industrial Health and Safety
Raw Materials/Monomers
Polymers/Polymerization
Compounding Ingredients
Intermediate and Semi-finished Products
Relating to Particular Industries and Fields of Use
Applications
Processing and Treatment
Properties and Testing
Book Reviews

Each issue has a subject index. Author and subject indexes are cumulated for each volume.

RAPRA Abstracts is available online through Pergamon Orbit Infoline: the computer file covers from 1972 onwards.

Plastics Abstracts

Published weekly by Plastics Investigations, Welwyn, Hertfordshire, UK, from 1959–83, this bulletin covered material relating to plastics and high polymers for Europe only. Each issue contained about 170 items.

Epoxy Resins and Plastics

Published monthly by R.H. Chandler Ltd., Braintree, Essex, UK, from 1980–83 this bulletin covered all aspects of epoxy resins. 200 items a month were selected from journals, books, technical reports, patents, conference proceedings, and trade literature.

Computer Databases and Databanks

In addition to abstracting publications, there are a number of computer data bases and data banks containing material relevant to polymers and plastics.

Some of these are summarized below:

Name:	RAPRA Tradenames
Type:	dictionary
Coverage:	Trade names/trade marks of rubber and plastic products and materials
Time Period:	1976–
Update:	twice monthly
Online host:	Pergamon Orbit Infoline

Name:	PLASPEC
Type:	data bank
Coverage:	Engineering, design, and processing characteristics of plastics. Specifications for plastics-processing machinery (North America only)
Time Period:	Current information
Update:	Weekly
Producer:	PLASPEC division of Plastics Technology Magazine
Online host:	ADP Network Services

Name:	PLASTISERV Plastics Information System
Type:	Suite of computer files
Coverage:	*Product Guide* — plastics available in N. America *SPI (Society of Plastics Industry)* — industry statistics; abstracts of technical and trade publications; legislation; user groups; SPI members and their services; SPI events *Buyers Guide* — product and service suppliers

	Industry Director — industry associations
	Wiley Catalog: catalogues covering plastics, polymers and chemicals
Time Period:	Varies by file
Update:	Varies by file
Producer:	DB Business Systems, Inc. and others
Online host:	DB Business Systems, Inc.

Name:	POLYCAS
Type:	dictionary
Coverage:	polymer and plastics nomenclature, trade names etc. taken from the CAS Registry Nomenclature file.
Time Period:	1965–
Update:	monthly
Producer:	CAS/Telesystemes-Questel
Online host:	Telesystemes-Questel (DARC)

Name:	IDC-Polymer-Speicher
Type:	data base
Coverage:	polymer and plastic patents
Time Period:	1976–
Update:	10,000 item/year
Producer:	IDC Internationale Dokumentationsgesellschaft fur Chemie mbH
Online host:	none — batch service only

Name:	Plastic Chemicals
Type:	data bank
Coverage:	economic/technocommercial statistics on plastics and chemicals for USA
Time Period:	varies; some from 1947–
Producer:	SAGE DATA Inc.
Update:	2–3 times weekly
Online host:	SAGE DATA Inc.

Name:	Plastics Chemicals Long Term Forecasts
Type:	data bank
Coverage:	economic forecasts for 12 key plastic chemicals for USA
Time Period:	1946–
Update:	annual
Producer:	Probe Economics, Inc.
Online host:	SAGE DATA, Inc.

Name: Plastics Chemicals Short Term Forecasts
Type: data bank
Coverage: economic forecasts for 12 key plastic chemicals for USA
Time Period: 1946–
Update: quarterly
Producer: Probe Economics Inc.
Online host: SAGE DATA Inc.

Name: CMAI Petrochemical Market Reports
Type: full-text/numeric
Coverage: economic statistics/forecasts for petrochemicals and plastic chemicals
Time Period: varies with material
Update: twice monthly or monthly, depending on material
Producer: Chemical Market Associates, Inc.
Online host: I. P. Sharp Associates

Name: Cordura Publications
Type: suite of data banks
Coverage: data banks on adhesives; elastomers; extruding and moulding plastics; plastic foams; and commercial names and sources in the plastics industry. Not available in Europe
Time Period: 1977– or 1978–, depending on subject
Producer: Cordura Publications

Name: GEISCO Computerized Plastics Retrieval System
Type: data bank
Coverage: products, properties, and vendors
Producer: GEISCO

Other abstracting publications and/or computer data bases containing polymer-related information are:

Name: Chemical Hazards in Industry
Type: publications/data base
Coverage: physical, chemical, and biological hazards in chemical and allied industries, waste disposal, health effects, legislation
Time Period: 1984–
Update: 250 items/month

Producer: Royal Society of Chemistry
Online hosts: Data-Star, Pergamon Orbit Infoline, STN

Name: Chemical Engineering Abstracts
Type: publication/data base
Coverage: chemical and process engineering
Time Period: 1971–
Update: 400 items/month
Producer: Royal Society of Chemistry
Online hosts: Data-Star, Pergamon Orbit Infoline, ESA-IRS

Name: Chemical Business NewsBase (CBNB)
Type: data base (subsets available in published form)
Coverage: chemical business information including grey literature (stockbroker reports, press releases, etc.)
Time Period: 1985–
Producer: Royal Society of Chemistry
Online hosts: Dialog, Data-Star, Pergamon Orbit Infoline, Finsbury Data Services

CHAPTER FOUR

Books, encyclopaedias and reviews

J. STUBBINGTON

Introduction

The phenomenal post-war growth of journal literature is paralleled in the production of equally large numbers of books, encyclopaedias and reviews. This inevitably limits the choice of what is nevertheless hoped will be representative samples of each category. A glance at the current publications catalogues of Butterworths, Elsevier, The Plastics and Rubber Institute and RAPRA Technology Ltd. to mention but a few publishers in this field reveals the target readership of such sources of information to be both numerous and specialised. The range extends from classroom textbooks for students — introducing fundamental principles — to authoritative review series and standard reference sources providing support across a wide spectrum of academic, industrial and commercial activities.

Books

Deservedly, books have gained recognition as the most important type of secondary publication, offering a more complete and better organized, albeit more succinct view of the subject matter than the undigested, widely dispersed originals. Delays inherent in the production of books, tend to blunt the force of novelty. Several years gestation may be required even with the advent of information technology into the publishing domain.

It is not really the function of the book to be the vehicle of first disclosure, rather it is to give perspective and depth to the perception of the topic. Books have a lasting value justifying the inclusion here of some earlier works.

GENERAL TEXTS

Some of the more popular titles include the *Textbook of polymer science* (1971, 2nd edn.), by F.W. Billmeyer Jnr (Wiley Interscience, Chichester). The first edition was published in 1962 as an undergraduate reader in the physical and organic chemistry of polymers and in instrumental analysis, but it will also appeal to the postgraduate polymer scientist both as a textbook and as a well-indexed reference source. The book is divided into five sections introducing the concept and characteristics of macromolecules, the structure and properties of bulk polymers and crystallinity, polymerisation kinetics (ionic and co-ordination), commercial polymers, their production, structure, properties, fabrication and uses, and polymer processing.

Macromolecules: an introduction to polymer science eds. F.A. Bovey and F.H. Winslow (Academic Press Inc., London, 1979). Aimed at the users and adapters of polymeric materials, this book was originally designed to be an introductory course in polymer science run by Bell Laboratories. All of the authors are employees of Bell and the emphasis is more on theoretical science rather than technological aspects, and is suitable for undergraduate or first year postgraduate courses. Coverage includes polymerisation kinetics, biological molecules, chemical reactions and degradation of polymers and the rotational isomeric state method of calculation of polymer chain dimensions.

Polymer handbook (2nd edn) eds. S. Brandrup and E.M. Immergut in collaboration with W. McDowell (Wiley-Interscience, John Wiley and Sons, London, 1975). The book provides a range of data and constants (with the exception of spectroscopic data) useful in both experimental and theoretical studies. Coverage is, however, restricted to synthetic polymers, polysaccharides and their derivatives. Physical and chemical constants of the laws of physics describing properties and behaviour are presented within reasonable or predictable limits. Constants dependent upon particular processing conditions or sample histories are not compiled, neither is there any attempt at critical evaluation of reported data. Polymerization, depolymerization, solid state and solution properties of commercially important polymers, physical data on oligomers, monomers, solvents, recent thermoplastics and a useful table containing practical data on commercial polymers are presented.

Plastics and rubbers: engineering design and applications by R.J.

Crawford (Mechanical Engineering Publications Ltd., London, 1985). The style is very readable and authoritative and the work provides a good grounding in the uses, possibilities and limitations of plastics and rubbers of relevance to both the specialist and the newcomer to processing, fabrication and design, well illustrated with case studies and supported with an extensive reading list.

Polymer materials: an introduction for technologists and scientists by C. Hall (Macmillan Press Ltd., London, 1981). The industrial production, and properties of commercial plastics, rubbers, adhesives, coatings and fibres are discussed covering molecular structure and morphology, mechanical, thermal, optical and chemical properties, including sections on permeability, degradation and flammability. Polymer nomenclature is discussed and a useful bibliography appended.

Handbook of plastics and elastomers by C.A. Harper (editor-in-chief) (McGraw-Hill Book Co., London, 1975). Although constrained by space considerations this book is a useful reference source for technological and applications data, with information on costs (although surely now of little relevance), design, fabrication and performance limits. Cross-references to appropriate standards and specifications are given, the work is extensively indexed and contains a separate glossary. Features covered include electrical, mechanical, chemical and environmental properties of laminates, reinforced plastics, composites, end-uses of man-made fibres, plastic and elastomeric foams, liquid and low pressure resin systems, plastic and elastomeric coatings and adhesives.

Molecular behaviour and the development of polymeric materials eds. A. Ledwith and A.M. North (Chapman and Hall, London 1975). The nature of ionic polymerisation and the role of macromolecular complexes in determining the course of polymerization reactions are covered and the dependence of macroscopic, physical and mechanical properties on fundamental molecular behaviour considered.

Polymer technology by D.C. Miles and J.H. Briston (Chemical Publishing Co. Inc., New York, 1979). The book was conceived as a primer for Plastics Institute, City and Guilds and Royal Institute of Chemistry examinations and covers thermosets, thermoplastics, elastomers, inorganic polymers, compounding ingredients, thermosetting and testing.

Fibres, films, plastics and rubbers: a handbook of common polymers by W.J. Roff and J.R. Scott with the assistance of J. Pacitti (Butterworths, London, 1971). This reference work is a classic of its kind, with tabulated data of value for both industrial process development and university laboratory. Natural and synthetic polymers are covered in the first 42 sections while the remaining 38 consider specific properties.

Polymer yearbook (2nd edn) (Harwood Academic Publishers, London, 1985). Although it does not claim to be fully comprehensive this interesting and valuable publication is an extremely useful desk reference tool. Future editions will be edited by an International Board of Editors which will ensure wider coverage, especially of Russian, American and Japanese sources. The reviews cover a wide range of topics, and are supplemented by a detailed glossary, tables, rules and formulae.

Rubber technology: natural rubber and the synthetics by P.W. Allen (Cosby Lockwood, London 1972). Technical and economic performance of the two kinds of rubber are highlighted and contrasted with reference to the structures of the producing and consuming industries. The basis for materials selection for particular applications is considered and future trends discussed.

RUBBER

Synthetic rubbers: their chemistry and technology by D.C. Blackley (Applied Science Publishers, London, 1983). Monomer sources, polymerization techniques and the effect of polymerization variables are considered for the main commercially available rubbers. Compounding, with especial reference to vulcanizing, reinforcement, stabilization against degradation and where appropriate, plasticization, and potential applications are discussed. The information presented tends to the broad generalization of widespread validity, rather than tabulations of specific data. Somewhat surprisingly, synthetic rubber latices appear to be outside the scope of the book.

Rubber technology and manufacture (3rd edn) ed. C.M. Blow, (Butterworths, London, 1982). Published in conjunction with the Plastics and Rubber Institute the book refers to the history of the rubber industry, focusing on the physics of raw and vulcanized rubbers, sources of raw polymeric materials, the chemistry and technology of vulcanization, compounding and reinforcing and filling materials, manufacturing and testing. Reference is made to the relevant standards. The book is well indexed, cross-referenced and includes a useful bibliography.

Kautschuk-Handbuch ed. S. Bostrom (Berliner Union, Stuttgart, 1959–62, German). Published in five volumes with a supplement, this is possibly one of the most comprehensive works on rubber technology produced, superseding the classic two volume *Handbook of rubber* (Hauser, 1935). The five volumes cover raw materials, synthetic rubbers, rubber compounding, textiles and machinery, rubber technology, machinery, mechanical goods, toys, cellular rubbers and manufactured goods, rubber products, latex technology and physical and mechanical testing of rubber. Volume five contains the cumulative index, and the supplementary volume is entitled *Techniques of measuring and control in the rubber industry* by J. Lerner (1961).

Developments with natural rubber ed. J.A. Brydson (Maclaren and Sons Ltd., London, 1967). Based on a symposium organized by the National College of Rubber Technology, this valuable survey discusses the developments within the natural rubber industry to counter the challenge of the newer synthetic materials. Subject areas considered include economics and marketing, production, oil-extension, preservation of vulcanizates, vulcanization, compounding, assessment of wear and skid-resistance, and uses of natural rubber with an assessment of its limitations.

Rubber technology: a basic course by A.S. Craig (Oliver and Boyd, London, 1963). Intended as an elementary textbook on the subject covering the City and Guilds Rubber Technology course of the period (including examination questions for revision purposes), the work nevertheless remains a useful general introduction to the subject. The presentation is informal and the basics are easy to grasp from this approach. The book covers historical aspects, preparation of raw rubber, synthetic rubber and rubber derivatives, compounding ingredients, the manufacture of rubber products and testing.

Introduction to rubber by J. Le Bas (transl. from French by J.H. Goundry) (MacLaren and Sons Ltd., London, 1965). Although short, this book is another useful elementary introduction to the subject covering the history of natural and synthetic rubbers, technological and industrial aspects.

Synthetic rubber technology, Vol. 1. Compounding, processing and application of standard types by W.S. Penn (MacLaren and Sons Ltd., London, 1960). This is a useful, student-level work with a practical bias. The different types of rubber, e.g. SBR, high-styrene resins, butyl, neoprene, nitrile, silicone rubbers and Thiokol are discussed. *Vol. 2* covers the newer types, with the exception of hard rubber, and sponge rubbers. Criteria for materials selection for different purposes, their compounding, processing and applications are discussed.

Rubber: natural and synthetic (2nd edn) by H.J. Stern (MacLaren and Sons Ltd., London, 1967). Following the first edition of 1954, the work is intended for users without specialised knowledge of rubbers. E.g. electrical, mechanical and civil engineers particularly, rubber converters and suppliers of chemicals and machinery to the industry should find the book of interest.

The Vanderbilt rubber handbook (12th edn) ed. R.O. Babbit (R.T. Vanderbilt Co. Inc., 1978). Although it may be considered as an extended piece of trade literature, the publication nevertheless deserves the status of a monograph. Emphasis is understandably on the use of Vanderbilt materials in the course of which the principles of successful compounding of rubber are thoroughly covered.

Bayer handbook for the rubber industry ed. S. Koch (Farbenfabriken Bayer AG, Leverkusen, 1970). Following on the success of the *Bayer pocket book for the rubber industry*, the *Handbook* was produced updating the physical, chemical and technological data relating to Bayer and associated companies, Stereokautschukwerke GmbH & Co. AG, Rhein-Chemie Rheinau GmbH and Haarmann & Reimer GmbH. The manual comprises two main sections, *Polymers* and *Chemicals*, with two short appendices, *Notes on the toxicology of rubber chemicals* and *Products supplied by Rhein-Chemie Rheinau GmbH*. At the end of each chapter are references to published and unpublished company studies, available directly from the company.

PLASTICS

Plastics technology: plastics materials, properties and applications by A.W. Birley and M.J. Scott (Leonard Hill, Glasgow, 1982). The choice of the right material at the right costs for a particular application or design requires the accurate assessment of many inter-related factors before the decision can be made. It is most helpful, therefore, to have a compact source at hand of carefully assessed basic data extracted from suppliers' product literature. General properties, processing techniques and design fundamentals are discussed, followed by reviews of the major polymer classes available from the main plastics raw materials manufacturers.

Plastics materials (4th edn) by J.A. Brydson (Butterworth Scientific, London, 1982). If there are gaps rising from the relationship between theoretical aspects of polymer science and technological practice and applications of plastics materials this book is a creditable attempt to bridge them. Properties of plastics materials are considered in the light of their composition and from the viewpoint of the molecular structure of the polymer. Examples of their performance in specific applications are also given. Elastomers and the main polymer types used for adhesives, fibres and coatings are discussed together with representative production or consumption statistics.

Plastics chemistry and technology by W.E. Driver (Van Nostrand Reinhold Co., New York, 1979). For anyone with only a limited background knowledge of chemistry this overview covers the ground from design to finished product, with an appraisal of polymerisation mechanisms, polymer 'idiosyncracies' and polymer interactions with the environment. Processing techniques, fillers, reinforcements, adhesion, test, methods and design concepts are also considered.

Plastics (6th edn) by J.H. Dubois and F.W. John (Van Nostrand Reinhold Co., New York, 1981). The first edition of this classic reference work was written before World War 2. Each new edition has continued the objective of achieving clarity of explanation and expression of complex concepts in chemistry, rheology and thermodynamics. Plastics materials data, processing and design methods, and the markets for the products are discussed with reference to a series of 'how-to' pointers on extrusion and injection moulding. Other processes covered include compression, transfer and injection moulding of thermosets, blow moulding and reinforced plastics fabrication. Also of interest are a materials selector, cost estimating, maintainance, heating and machine control problem-solving. A useful glossary and acronym list is provided.

Plastics engineering handbook of the Society of the Plastics Industry Inc. (4th edn) ed. J. Frados (Van Nostrand Reinhold Co., New York, 1976). First published in 1947, this handbook features much of interest to scientists, technologists, administrators, and marketing and sales personnel. Coverage ranges from the selection of materials, processing, finishing, to ultimate uses, and limitations restricting their use are discussed. The basic chemistry of polymers, their properties and applications, are complemented by sections on injection moulding, extrusion, thermosetting, thermoforming, blow moulding, rotational moulding, calendering, processing of vinyl dispersions, powder coating, casting, machinery and equipment. Also useful are sections on reinforced and cellular plastics, radiation processing, mould-making and materials, the design of moulded products, standards for tolerances of mouldings, compounding, materials handling and performance testing of mouldings. A good index and glossary round off this useful book.

The plastics engineers' data book by A.B. Glanvill (The Machinery Publishing Co. Ltd., Brighton, 1971). Basic processing and product data are tabulated or presented graphically, and the methods of calculation are detailed, for the major plastics converting processes, injection moulding, blow moulding, extrusion, thermoforming, thermoset moulding, film and extrusion coating. Also of interest are design data for tools and dies, processing data and product testing methods. Growth trends for all the main plastics-forming methods are given.

Engineering with polymers by P.C. Powell (Chapman and Hall, London, 1983). This is another useful book on aspects of designing with plastics which discusses the nature and behaviour of polymers in the light of standard engineering techniques of stress analysis, structures, fluid mechanics, and heat transfer.

Technicians' handbook of plastics by P.A. Grandilli (Van Nostrand Reinhold Co., New York, 1981). Plastics production studies by leading manufacturers are reported, intended to be used as on-the-job reference material which is both broad in scope and with a concise and readable style. Particularly useful are sections on polymer basics, nomenclature, shop jargon and the extensive glossary. Plastics applications in electronics, thermoset laboratories, techniques for working with graphite and aromatic polyamide (Kevlar) fibres and determination of the resin content of such laminates are described. Safe working procedures for laboratories are emphasized.

Plastics technology by R.V. Milby (McGraw-Hill Book Co., New York, 1973). Based on extensive experience of the industry and of technician training, the author covers step-by-step, both the theory and practice of thermoset and thermoplastic material processing and testing. Lists of trade names, professional groups and other organisations supplement the main text.

Plastics technology: theory, design and manufacture by W.J. Patton (Reston Publishing Co. Inc., Reston, 1976). The nature of plastics materials and the individual 'personality' of each type is described, and basic theories of polymer chemistry and successful design methods are explained, with reference to normal and unusual properties of the materials, selection criteria and manufacturing. Cellular plastics, and fillers, reinforcements and additives are covered in some depth.

Kunststoff taschenbuch (international plastics handbook) by H. Saechtling (English transl. by M. Kaufman) (Carl Hanser Verlag, Munchen, 1983). Over 20 editions of this well known work, sponsored by the Society of Plastics Engineers, have appeared since it was first published in 1936. The handbook contains much useful reference material, an international trade names directory, and lists German plastics textbooks and standard abbreviations used in the plastics industry. Details of current commercial developments, processing techniques, materials and their testing are presented.

Plastics materials and processes by S.S. Schwarz and S.H. Goodman (Van Nostrand Reinhold Co., New York, 1982). Chemistry, properties and applications of plastics, including high molecular weight polyethylene, fluorocarbon copolymers, thermoplastic polyesters and polyacrylates, thermoplastic elastomers and high temperature resistant polymers are covered and general purpose, engineering and speciality thermoplastics are thoroughly analysed. Aspects of reinforced plastics are also covered. Historical development of processes lead to discussion of the capabilities and limitations of e.g co-injection and co-extrusion processes, SMC, TMC, RTM, ERM, injection moulding and RIM, structural foam moulding, joining and fastening techniques, decorating and finishing and electronic process control. Safety procedures and criteria for materials and process selection are discussed. A useful list of trade names supplements the main text.

Handbook of plastics (2nd edn) by H.R. Simonds, A.J. Weith and M.H. Bigelow (Van Nostrand Co. Inc., London, 1949). For those interested in the development of the plastics industry during World War 2, a period of burgeoning growth, the properties, manufacture, fabrication and applications of plastics available at that time are described. Polymer chemistry, analysis, post-forming, low-pressure moulding techniques and laminating are covered. There are also sections on patents, costing, a trade names list, and a technical and chemical glossary.

Plastics process engineering by J.L. Throne (Marcel Dekker Inc., New York, 1979). Senior undergraduates and first year graduates in chemical and mechanical engineering intending to specialize in plastics processing are addressed. Fundamentals of reaction kinetics, basic concepts of heat transfer, solution thermodynamics and major processing techniques and their underlying principles are considered. Newer developments in thermoplastics, structural foam moulding and injection blow moulding are referred to as examples of new technology. The work is extensively indexed, cross-referenced and supplemented by 700 bibliographical references and 400 'homework' problems!

Plastics 1980: desktop data bank (books A and B) (4th edn) (International Plastics Selector Inc., a subsidiary of Cordura Publications Inc., 1979). *Book A* lists the classes of materials and their properties, while *Book B* is concerned with extrusion and moulding grades of plastics with details of properties, including flammability and creep. The work is useful as a materials selector although data is reproduced only from the manufacturers' literature available to the compilers and the coverage is therefore not comprehensive. Indexes to the materials and lists of manufacturers and suppliers are included.

Polymer chemistry: polymers: chemistry and physics of modern materials by J.M.G. Cowie (International Textbook Co. Ltd., Aylesbury, 1973). In attempting to give as complete a picture as possible of polymer science, preparation, characterization and properties and applications are considered. The main features and kinetics of polymerization reactions, the structure-property relationships of stereoregular polymers and copolymers and their control are discussed. Polymer characterisation includes sections on dilute solution behaviour and crystalline, amorphous and elastomeric states. Mechanical properties of the polymers are related to their ultimate use. Stability and degradation (those largely synonymous terms for the same phenomena) are not covered to any extent, an important omission in a book of this type.

Chemistry of natural and synthetic rubbers by H.L. Fisher (Reinhold Publishing Corp., New York, 1957). With due allowance for its age this is a useful source of historical data on natural and synthetic rubbers, with discussion on vulcanization, acceleration, anti-oxidation and anti-ozonation. Natural rubber, latex, synthetic rubber, hard rubber and reclaimed rubber are considered, together with the more important properties of these. Raw materials for synthetics and chemical derivatives, and the bonding of rubber to metal are also examined.

Principles of polymer chemistry by P.J. Flory (Cornell University Press, Ithaca, New York, 1953). A classic work by an acknowledged pioneer in polymer science, Flory draws heavily on experimental work for illustration of specific polymer properties and characterization methods. Early developments and fundamental concepts in polymerization and copolymerization, reaction mechanisms and kinetics, polymer constitution, structure, molecular weight determination and distribution, molecular configuration and associated properties of polymers and their solution gives a useful reference to the state of the art in the early 1950s. A glossary of widely used symbols is included.

Polymer chemistry by D.B.V. Parker (Applied Science Publishers Ltd., Barking, 1974). Characteristics of monomers and macromolecules, the principles of polymerization and polycondensation and the importance of kinetics in manufacturing processes are discussed. The influence of polymer morphology, the arrangement of the monomer 'building brick' on polymer properties, the effects of molecular weight on the behaviour of polydisperse solutions and degradation mechanisms are considered.

High polymeric chemistry by W.S. Penn (Chapman and Hall Ltd., London, 1949). This work remains a useful reference source for theoretical polymer chemistry and supporting practical methods. Hydrocarbon-substituted polyethylenes, vinyl polymers, polydienes from hydrocarbons and halogenated hydrocarbons are covered.

Organic polymer chemistry: an introduction to organic chemistry of adhesives, fibres, paints, plastics and rubbers by K.J. Saunders (Chapman and Hall Ltd., London, 1973). Theoretical aspects of polymer chemistry are related to industrial practice in the production and use of commercially significant polymers.

Monograph series

Several of the larger technical publishing houses and the major technical and research organisations working within the field of polymers and plastics have published some excellent series. The following are well worth examination, although the choice has been selective owing to the wealth of material available.

Contemporary topics in polymer science (Plenum Publishing Corp.) *Volume 1: macromolecular science: retrospect and prospect* ed. R.D. Ulrich, 1978. The history of modern theories of chemistry, morphology, physics and rheology of polymers, current research and predictions for future developments are outlined.

Volume 2 eds. E.M. Pearce and J.R. Schaefgen, 1977. This volume contains reports on the latest investigations of polymeric liquid crystal systems, polyblends, novel polymers, conformational transitions, rubber elasticity theory and synthetic procedures for stiffening polymer chains.

Volume 3 ed. M. Shen 1979. Volume 3 is divided into three parts which cover polymer synthesis, characterisation and clustering phenomena in ionic polymers, microdomain formation in block copolymers, glass transition phenomena, highly oriented polymers and gel entrapment systems.

Volume 4 eds. W.J. Bailey and T. Tsuruta, 1984. Volume 4 includes 63 papers from leading polymer chemists in the USA and Japan. Work described includes gas-phase studies, copolymerisation of carbon dioxide and epoxides, effects of branching and solution concentration on polymer chain dynamics and modification of polymer surfaces by surface active graft copolymers.

Volume 5 ed. E.J. Vandenberg, 1984. Particular topics of interest covered in Volume 5 include biochemical and high temperature polymers, their synthesis, properties and applications.

Polymer science and technology (Plenum Press). This series covers the latest research on the structural, mechanical, morphological and technical aspects of polymers.

Plastics institute monographs (Iliffe Books Ltd., London, 1946–74). This series is currently published by the Plastics and Rubber Institute from 1975 onwards.

Society of Plastics Engineers monographs (John Wiley and Sons, New York). Designed to promote polymer science and engineering the series covers *Injection moulding*, ed. I.I. Rubin; *Nylon plastics*, ed. M.I. Kohan; *Introduction to polymer science and technology: an SPE textbook*, eds. M.S. Kaufman and J.J.

Falcetta; *Principles of polymer processing*, eds. Z. Tadmoor and C.G. Gogos; *Colouring of plastics*, eds. T.G. Webber; *The technology of plasticisers*, eds. J.K. Sears and J.R. Darby; *Fundamental principles of polymeric materials*, ed. S.L. Rosen; and *Plastics polymer science and technology*, ed. M.D. Baijal.

Natural Rubber Producers' Research Association monograph series (MacLaren and Sons Ltd., London). This series includes titles such as *The chemistry and physics of rubber-like substances* (L. Bateman, 1963).

Chemistry in modern industry series — plastic and rubbers eds. E.W. Dick and D.J. Daniels (Butterworths, London, 1971). This work is a concise introduction to polymer science and the manufacture and processing of polymeric materials. Particular emphasis is given to processing techniques.

Ellis Horwood Series in chemical science (Ellis Horwood, Chichester). This series includes volumes of particular interest to the industrial polymer scientist, e.g. *Kinetics and mechanisms of polymerisation reactions: applications and physiochemical systematics*, by P.E.M. Allen and C.R. Patrick; *Polymers and their properties*, by J.W.S. Hearle; *Liquid crystals and plastic crystals*, by W. Gray and A. Winsor.

Development Series (Elsevier Applied Science Publishers Ltd., London). This series attempts to recognize and expand on rapidly occurring developments, events important enough to place in context as they occur, a function also of the journal literature but often less effectively carried out owing to the scattering of the information over many sources. Volumes appearing at regular intervals include: *Developments in plastics technology*, eds. A. Whelan and J.L. Craft; *Developments in plastics and rubber composites*, ed. C.W. Evans; *Developments in rubber technology*, eds. A. Whelan and K.S. Lee; *Developments in polymer characterisation*, ed. J.V. Dawkins; *Developments in block copolymers*, ed. I. Goodman; *Developments in polymer degradation*, ed. N. Grassie; *Developments in GRP technology*, ed. B. Harris; *Developments in polymerisation*, ed. R.N. Haward; *Developments in reinforced plastics*, ed. G. Pritchard; *Developments in polymer stabilisation*, ed. G. Scott; *Developments in injection moulding*, eds. A. Whelan and J. Goff; *Developments in ionic polymers*, eds. A.D. Wilson and H.J. Prosser; *Developments in polymer photochemistry*, ed. N.S. Allen; *Developments in oriented polymers*, ed. I.M. Ward; *Developments in crystalline polymers*, ed. D.C. Bassett; *Developments in adhesives*, ed. A.J. Kinloch; *Developments in PVC production and processing*; eds. A.Whelan and J.L. Craft; *Developments with thermosetting plastics*, eds. A. Whelan and J.A. Brydson.

The Plastics and Rubber Institute series of monographs ranges widely over materials, processing technology, properties and testing. The current catalogue includes *Plastics films* (2nd edn) by J.M. Briston (1983), covering film forming polymers, their manufacture, properties, conversion and application; *Injection mould design*, (3rd edn) by R.G.W. Pye (1983); *Thermosetting plastics: practical moulding technology*, ed. J.F. Monk (1981); *Injection moulding of engineering thermoplastics*, by A. Whelan and J. Goff (1987); *Flow properties of plastics melts* (2nd edn), by J.A. Brydson (1981); and its companion, *Polymer melt rheology* by F.N. Cogswell (1981). Guidance on the testing of plastics and the interpretation of the results is given in *Mechanical testing of plastics* (2nd edn) by S. Turner (1984) again with a cognate work *Handbook of plastics test methods* (2nd edn) by R.P. Brown (1981), covering physical test procedures dealing with international and national standards.

The American Chemical Society series of scientific and technical monographs (Rubber Division, American Chemical Society) has a number of titles of particular interest including *The chemistry and technology of rubber*, eds. C.C. Davies and J.T. Blake, (1937, Reinhold Publishing Corp.); *Science and technology of rubber*, ed. F.R. Finch (Pandemic Press, 1978); *Synthetic rubber*, eds. G.S. Whitby, C.C. Davies and R.F. Dunbrook (1954, John Wiley and Sons Inc., New York); *Rubber technology* (2nd edn) ed. M.Morton (Van Nostrand Rheinhold Co. Inc., 1973).

Encyclopaedias

Encyclopaedias offer a reliable, authoritative and rapid lead-in to subjects which may be unfamiliar, covering a wide range of topics which may be highly specialized, e.g. individual polymers or polymer classes, properties, processes and uses. They tend to be extremely expensive to produce, although the cost may be spread over a long period of time, corresponding to the appearance of individual volumes. Cumulative indexes are inevitably the last to appear, although individual volumes may be indexed. However, they form an essential source of reference material which may otherwise not be immediately obtainable, online data base searching notwithstanding, and frequently furnish the ideal starting point for seeking the answer to the problems.

Encyclopaedia of polymer science and technology: plastics, resins, rubbers, fibres ed. N.M. Bikales (New York, John Wiley and Sons, 1964–72, 16 volumes, plus supplement). Contributions submitted by specialists worldwide, cover a wide range of monomers, polymers and their properties, methods of synthesis and commercial production processes are reviewed. Practical and theoretical aspects are discussed in depth, and cross-referencing is consistently good enough to suggest further leads to follow. Extensive bibliographies of literature references and patents, where appropriate, are appended.

Encyclopaedia of polymer science and engineering (2nd edn) by J.I. Kroschwitz (editor-in-chief) (New York, John Wiley and Sons) was started in 1985 and although not yet complete, this is a very substantial work indeed being a completely revised version of the classic *Encyclopedia of polymer science and technology*, started in 1964. An interesting feature of the current work is the bibliography of papers published by Herman Mark in honour of his 90th birthday.

Modern plastics encyclopaedia (McGraw-Hill Publishing Co., New York) is published annually as part of *Modern plastics*. There are four sections: *Textbook*, presenting general review articles on materials and processes; *Design guide*, property guide and materials selector; *Databank*, giving design and specification data on materials and equipment; *Directory of Suppliers*, providing classified lists of over 4000 companies, mostly situated in the USA, producing materials, machinery and services.

The encyclopedia of PVC is a four-volume work (Marcel Dekker, New York and Basel, 1986) eds. L.I. Ling and C.A. Heiberger. The four volumes cover resin manufacture and properties, compound design and additives, compounding processes, product design and specifications.

Non-English language encyclopaedias include the annual *Encyclopédie française des matières plastiques* (Les Publ. Techniques Assocn., Paris) and the *Enciclopedia — Annuario delle Materie Plastiche* (Materie Plastiche ed Elastomeri, Milan, Annual, 2 volumes).

Another English language encyclopaedia of wider scope than the plastics and polymers field is the *Kirk-Othmer concise encyclopaedia of chemical technology* ed. M. Grayson (John Wiley and Sons, New York, 1985). Based on its wide-ranging counterpart, the 24 volume (plus supplement) third edition of the *Kirk-Othmer encyclopedia of chemical technology*, this single massive volume is an abridgement of very high quality. The format is modelled on that of the parent publication, with subject coverage ranging from acrylamide to Ziegler-Natta catalysis. Copiously cross-referenced and with a good keyterm index, this work is a valuable addition to collections of instantly available reference material.

Reviews

In many respects, review articles resemble entries in encyclopaedias, but with the advantages of lower cost, more frequent updating, and potential for greater topic specialisation within the series. Some monographs might properly be described as reviews, but the distinction is not essential provided that the indexing is adequate and the treatment authoritative. Although it is not unknown for fresh lines of research to be undertaken without reference to previous literature, a review often represents a good lead-in to the subject, particularly when it is critically assessed. It was with precisely this aim in mind, of minimizing the need for labour-intensive literature search and evaluation that the *Index to reviews in organic chemistry*, was initiated by D.A. Lewis at ICI (Plastics Division). It was later published by the Royal Society of Chemistry, having proved its value over the years, but unfortunately ceased publication in 1983.

A cognate work, no longer published, *Chemical Abstracts review index (CARI)*, is a comprehensive KWIC index to review articles appearing in each volume of *Chemical Abstracts*, giving rapid access to the abstract number. KWIC is not the easiest of formats to read, as each variation of the keyword in the titles is listed which may result in separation of related material, e.g. POLYMER is separated from POLYMER(S) by POLYMERASE(S), POLYMERIC(S), POLYMERIZATION(S), POLYMERIZED and POLYMERIZING. Some persistence is needed to scan the columns to be sure nothing is overlooked.

Recent developments in polymer technology is a publication started in 1987 by RAPRA Technology Ltd., as an annual series of reviews of trends in the rubber and plastics industry, supplemented with extensive bibliographies.

RAPRA review reports ed. C.A. Green is a quarterly journal published by Pergamon Press Ltd., Oxford, containing three articles in each issue compiled by acknowledged leaders in each field and accompanied by a comprehensive bibliography extracted from the *RAPRA Abstracts* online service.

A further review by RAPRA Technology Ltd. entitled *1000 key reviews in polymer technology*, ed. R. van den Hondel (1987) covers reviews published since 1982 on manufacture, process and applications for plastics, rubber and composites.

The Plastics and Rubber Institute (PRI) also publish a series of technical reviews including titles such as *Design factors for unreinforced thermoplastics products with specific reference to pipes* (1976); *Design factors for unreinforced thermoplastics products with specific reference to containers* and *Multiple transitions in semi-crystalline polymers* by R.F. Boyer, (1973).

An updated version of polymer research projects in UK universities and research institutes is to be found in the *7th review of research activities in polymer science and technology* (1986), giving details of research personnel responsible for the project and accompanying test facilities.

A joint publication of the PRI and RAPRA Technology Ltd. is the quarterly journal *Progress in rubber and plastics technology*. Topics covered in 1987–88 include quality control in the rubber industry, acrylic elastomers, new vulcanizing systems, injection moulding of fibre-reinforced plastics, polymer blends, polymeric grid materials for civil engineering and construction, polymeric materials for semi-conductors and project management.

CHAPTER FIVE

Patents and trade marks

J. F. SIBLEY

Introduction

A trade mark is a means of identification which may be one or more words or a device or a combination of these. It is used by a trader to make his goods or services readily distinguishable from others in the market place. Many, but not all trade marks are registered.

A search can be made at the national Trade Marks Registry to determine the proprietor and nature of goods or services covered by a registered trade mark. Today, online data bases are available which contain data on International, United Kingdom, United States and French trade marks.

Sometimes the patent protecting a particular product may be found by searching the Chemical Abstracts Registry File for the trade mark or trade name but, in general, trade marks are of little value as a source of technical information since they can be used for a variety of similar products the nature of which can change to suit market conditions.

On the other hand, the patent specifications published throughout the World constitute an important and extensive information source. At the beginning of 1987, patent specifications were being published at the rate of more than 750,000 a year by more than 50 patent issuing offices. Derwent Publications Ltd., London, process most of these specifications for inclusion in their *World patents index*, a comprehensive patent abstracting and indexing service covering the whole range of technology.

Section A of *World patents index* (WPI) is titled PLASDOC and this is devoted to patent specifications covering inventions relevant to the plastics industry in the broadest sense. The section covers polymers and resins, polymerisation processes and equipment, fabrication processes and equipment, monomers and monomer production and purification, additives, polymerisation catalysts, and uses and applications of polymers and resins.

Each week, full details of about 950 new inventions are added to the PLASDOC section of WPI; this means that about 50,000 patents are published each year which contain new technical information of interest to the polymer and plastics industry.

Various studies have shown that most of the information contained in patent specifications is never published elsewhere and, even when it is, it is usually some time after the publication of the specification. Patent systems have a long history and the total number of specifications published to date exceeds 25 million.

Patent Systems

Most countries operate a patent system in which an applicant is given monopoly rights in the use of his invention for a limited period in return for disclosing the nature of that invention in sufficient detail for it to contribute to the stock of public knowledge.

The various national patent systems have the same basic intention, but distinct variations developed not only in the period of the monopoly granted, but also in the nature and concept of invention. All national systems involve submission of a patent specification to the national patent office where it remains secret until publication which may occur after a fixed period of time or after acceptance or grant. Grant of a patent follows an examination of the specification by an examiner at the national patent office to ensure that it satisfies the requirements of the national patent act.

Since the protection offered by a patent is restricted to the country granting the monopoly, the applicant or his agent has to file or submit a patent application to the national patent office of every country in which protection is sought. The Paris Convention of 1883 facilitates this by allowing the applicant a period of one year from the date of filing his original national patent application in which to decide on the value of his invention and the extent of foreign patent protection required. The filing date of the first filed national application is known as the 'priority date' and subsequent foreign applications are often called 'convention' or 'equivalent' applications. Thus one invention can be protected internationally by a family of national patents.

Many patent issuing authorities have adopted an examination system where the specification is published 18 months after the priority date sometimes together with the search report giving the prior art found by the examiner. The applicant is then given a period of time in which he can consider the relevance of the prior art and decide whether or not he wants to continue with his application. If he does, he then has to apply for substantive examination in which the application is subjected to examination by the patent office to determine whether a patent should be granted. This system leads to multiple publication of each specification, first as a published unexamined application when the serial number is suffixed by a letter 'A' often with a digit, then secondly, if the applicant is successful in prosecuting the application, as a granted patent when the serial number is suffixed by the letter 'B'. In Japan, as used to occur in Germany, there is a three-stage publication procedure. In a few countries, for example the USA and Canada, an older system is adhered to in which pubication occurs only once, after grant of a patent.

Two fairly recent developments in the patent area have been the Patent Co-operation Treaty (PCT) and the European Patent Convention (EPC) which are intended to facilitate the procedure for obtaining patents in a number of countries which are contracting states of the EPC or PCT. In both systems, the applicant makes a single application to an appropriate receiving office designating the countries in which he seeks protection. The EPC covers the 13 contracting states within Europe, whereas the PCT covers a large number of countries in Europe, Africa, Asia and the Americas. The procedural steps to be followed differ in the two systems but both systems involve substantive examination, and successful applications lead eventually to a bundle of national patents which are equivalent to those which could have been obtained by individual national applications. EPC applications must be filed in or translated into one of the three official languages, French, German or English and, once a patent has been granted, the text of the claims of the patent must be provided in each of the three official languages. Many contracting states also require a translation of the granted patent specification to be filed in an official language of the state concerned, but the United Kingdom did not invoke this section of the Act until September 1987. This means that although all future granted EPC cases designating the United Kingdom will be in English, a number will exist in French or German.

A granted patent gives the proprietor the right to prevent others from making, using or selling the invention for as long as he keeps the patent in force or until it expires at the end of its term. It is important to remember that this right applies only in the country granting the patent. It is also important to realize that a significant number of patent applications do not lead to granted patents and that even where a patent is granted, the proprietor may not keep it in force for its full term if this requires the payment of quite substantial annual renewal fees. However, whilst the monopoly is limited, the value of the patent as a piece of literature is timeless and free from any geographical restriction.

Modern patent documents comprise three parts. The first is the cover page on which are printed the bibliographic details and usually an abstract (Fig. 5.1). The second is the specification itself which contains the description of the invention together with references to earlier work in the field and often sets out the reasoning which led to the present invention together with chemical formulae, drawings and examples as appropriate. The third part of the patent document comprises the claims which define precisely the scope of the monopoly sought.

Fig. 5.1 shows the front page of a published European patent application. Each data element such as publication number, application and priority details, inventor and applicant are numbered according to an agreed standard so that such data can be retrieved from any patent specification regardless of the language in which it is printed.

Information from patents

Patents specifications can provide three types of information; these are (1) information on monopoly rights, (2) commercial intelligence and (3) technical information. Finding or retrieving information from the patent literature is known as searching.

MONOPOLY RIGHTS

The existence and extent of third-party monopoly rights is of interest to the industrialist. The fact that a patent has been granted does not mean that the proprietor is able to practice his invention freely. It may be that prior patent rights exist which would be infringed in so doing. Therefore an infringement search should be carried out before commercialization of a product or process to find any third-party patents that are currently in force and which could obstruct such commercialization. Such searches are restricted to patents which could be in force, in general to patents published in the last 20 years, in the countries in which the industrialist wishes to operate. Should such patents be found, the official records are checked to determine whether these patents are still in

(10) Europälsches Patentamt
European Patent Office
Office européen des brevets

(11) Publication number: 0 197 310 A2

(12) EUROPEAN PATENT APPLICATION

(21) Application number: 86103014.6

(51) Int. Cl.[4]: C 08 F 10/00
C 08 F 4/64

(22) Date of filing: 06.03.86

(30) Priority: 07.03.85 JP 45662/85
18.03.85 JP 53849/85
19.03.83 JP 54838/85

(43) Date of publication of application:
15.10.86 Bulletin 86/42

(84) Designated Contracting States:
BE DE FR GB NL

(71) Applicant: MITSUBISHI PETROCHEMICAL CO., LTD.
5-2, 2-chome, Marunouchl
Chiyoda-ku Tokyo 100(JP)

(72) Inventor: Matsuura, Mitsuyukl
Mitsubishi Petrochemical Co. Ltd. Jushi Kenkyusho
1, Toho-Cho Yokkalchl-Shl Mle-Ken(JP)

(72) Inventor: Fujita, Takashi
Mitsubishi Petrochemical Co. Ltd. Jushi Kenkyusho
1, Toho-Cho Yokkaichl-Shl Mle-Ken(JP)

(74) Representative: Hassler, Werner, Dr.
Postfach 17 04 Assenberg 62
D.5880 LOdenscheld(DE)

(54) Catalyst components for polymerlzing olefins.

(57) A catalyst component for polymerizing olefins, characterized by a contact product of the following components (A_1), (A_2) and (A_3):

component (A_1): a contact product of, as essential components, a magnesium dihalide such as $MgCl_2$, a titanium tetraalkoxide such as $Ti(On\text{-}Bu)_4$ and/or a polymer therof, and a polymeric silicone compound having the structure represented by the formula:

$$-\underset{\displaystyle H}{\overset{\displaystyle R^1}{\underset{|}{\overset{|}{Si}}}}-O$$

(wherein R^1 denotes a hydrocarbon residue) such as methylhydropolysiloxane,

component (A_2): and acid halide compound such as $\begin{matrix} COCl \\ COCl \end{matrix}$ ortho-$C_6H_4(COCl)_2$, and

component (A_2): a liquid titanium compound such as $TiCl_4$ (component (a)) and/or a halogenosilicon compound such as $SiCl_4$ (component (b)), provided that the component (a) contains a halogen when the component (a) is used alone.

EP 0 197 310 A2

Croydon Printing Company Ltd

Figure 5.1

force. The searcher may then go on to try to find invalidating prior art by looking for relevant information published before the priority date of each patent. Novelty searches which are carried out to establish the patentability of an invention before filing are really extensions of state-of-the-art searches and can be considered as part of type (3).

COMMERCIAL INTELLIGENCE

Commercial intelligence is gaining in importance and many companies are trying to predict areas of emerging commercial technology. Since such technology can be expected to be patented, the patent literature can be used to highlight these developments at an early stage. Commercial software is available to help with such analyses. By studying the patent portfolios of competitors, a company may be able to detect new products or processes which might become important rivals in the market place or provide useful licensing opportunities in the future. The same sort of studies can be made before entering into takeover negotiations or looking for key personnel.

TECHNICAL INFORMATION

Technical information or subject matter searches are many and varied. Some state-of-the-art searches can be made with only a cursory study of the patent literature but, where the subject of a search is in a technical area, such searches, especially when carried out to determine novelty of an invention or to plan an area of commercial research, should include patent documents.

The problem facing the searcher is knowing how to start the search, what data bases to use and what classification systems are available. It is possible to use official patent collections directly or to start with one or more of the secondary abstracting services which cover patent specifications, turning to the full text of the specification as necessary. One group of classification systems that can be used are those used by patent offices themselves when carrying out searches to establish patentability. To make best use of these, the searcher needs to have a comprehensive understanding of the structure of the systems and the philosophy and principles behind its application by the offices.

Patent office classification systems

There are three systems which are readily available. The most widely used system is the International Patent Classification (IPC) which is employed by over 40 different patent issuing authorities to a varying extent. The other two are the domestic classification systems used by the United Kingdom and the United States Patent Offices in addtion to the IPC.

The International Patent Classification (IPC)

The IPC is intended to yield an internationally uniform classification of patent documents which can be used to establish the novelty and evaluate the inventive step of inventions claimed in patent applications.

The IPC is revised on a regular basis, at present at five-year intervals by an international committee of experts under the auspices of the World Intellectual Property Organisation (WIPO). It does not have a long history. The first edition of the IPC came into force in September, 1968; the second edition in July, 1974; the third edition in January, 1980 and the fourth and current edition in January, 1985.

The IPC divides the whole of technology into eight sections indicated by the capital letters A to H. Each section is in turn broken down into classes indicated by the section letter followed by a two-digit number and these classes are further divided into sub-classes by adding another capital letter to the class symbol. In turn, the sub-classes are divided into main groups and further into subgroups. This can be illustrated as follows.

Section C	— Chemistry; metallurgy
Class C08	— Organic macromolecular compounds; their preparation or chemical working-up; compositions based thereon
Subclass C08F	— Macromolecular compounds obtained by reactions only involving carbon-to-carbon unsaturated bonds
Main Group C08F-2/00	— Processes of polymerisation [2]
Subgroup C08F-2/02	— Polymerisation in bulk [2]

The [2] indicates that both the group and subgroup have been operative since the second edition of the IPC and can be used to search the patent literature since July 1974.

It is important to note when a particular group or subgroup

became operative because most formal patent data bases are not reclassified when new editions of the IPC are introduced.

The classification is hierarchical, the hierarchy among groups being determined solely by the number of dots preceding the titles of the subgroups and not by their numbering. Notes are associated with each section, class and subclass and these must be read carefully by the searcher if best use is to be made of the classification.

Copies of the various editions of the IPC are generally available for examination in the various libraries which keep patent documents and copies of the current edition can be purchased from Carl Heymanns Verlag KG, Munich.

The IPC is intended to ensure that any technical subject with which an invention is directly concerned can be classified as far as possible as a whole and not by the separate constituent parts.

Patent documents usually contain two types of technical information; firstly there is the technical information relating to the invention itself, which is called the 'invention information' and secondly there is the 'additional information'. This may be (a) non-trivial technical information which is not claimed and which does not form part of the invention but which might be of value in future patentability searches in this area or (b) technical information which complements information already classified.

The various patent issuing authorities are obliged to classify the 'invention information' but the application of futher IPC symbols to classify 'additional information' is not compulsory. The IPC symbols representing the 'invention information' are separated from those used to represent 'additional information' by a double oblique stroke. The European published application, EP 0197310 (Fig. 5.1) has two invention information symbols C08F 10/00 and C08F 4/64 from the fourth edition of the key. The first, given in bold type, is seen as the symbol which most adequately represents the invention, a feature which can assist when using the Pergamon Orbit InfoLine 'get' or the Telesystemes Questel 'memsort' command for patent statistics.

With the fourth edition of the IPC, hybrid classification was introduced in parts of the key. Here, additional IPC symbols are used as indexing terms to identify such things as essential constituents of a composition or mixture, or uses or applications of classified technical subject matter.

Domestic classifications

The United Kingdom classification, like the IPC, divides the whole of technology into eight sections indicated by the letters

A to H. Each section has a number of divisions indicated by the section letter and a number. In turn, these divisions are divided into a number of headings by the addition of another letter.

Thus we have:

Section C; Chemistry, metallurgy
C1 Inorganic chemistry; glass; fertilizers; explosives
C2 Organic chemistry
C3 Macromolecular compounds
C4 Dyes; paints; miscellaneous compositions

Division C3, is broken down further into the following headings:

C3 Macromolecular compounds
C3A Cellulose derivatives etc.
C3B Epoxy resins
C3C Cellular polymeric materials
C3E Indiarubber etc.
C3F Cellulose etc. compositions
C3H Proteins
C3K Polymers etc. additives
C3L Polymers etc. working up
C3M Polymer etc. blends
C3N Plastic compositions etc.
C3P Addition polymers etc.
C3Q Indiarubber etc. compositions
C3R Condensation polymers etc.
C3S Organic silicon compounds
C3T Organosilicon polymers
C3U Polysaccharides etc.
C3V Polymer etc. compositions
C3W Indexing schedule for polymer information

As with the IPC, it is essential that the searcher reads and understands the notes which are featured at the beginning of each heading and, often, within the body of the heading. This will, for example, explain the difference between C3N and C3V. The United Kingdom Patent Office has run training courses in the use of their classification.

The United Kingdom classification is regularly revised and classification keys are issued each year. The headings of most interest to the polymer and plastics industry are given above, documents within each heading being broken down using a series of classifying and indexing terms chosen from schedules given within the heading.

The Patent Office examiner will assign to a document, the classifying term which, in his opinion, most adequately represents the technical subject of the invention as a whole. Rarely will a document be given more than one classifying term. Indexing terms are assigned to a patent specification usually to reflect only part of the disclosure and usually without regard to novelty of the statement being indexed. In general, the classifying terms are broken down into groups and can only attract indexing terms from a restricted list.

Recently the United Kingdom Patent Office introduced a Universal Indexing Schedule for Use, Applications, Utility and Property. This is the UIS schedule which, although not truly universal in that it is only applied to about 60% of United Kingdom patent specifications, does provide a very useful entry into the patent literature in a way not previously available. Use of the schedule will make it easier for searchers whose interest lies in particular products or manufacturing operations or in particular properties or utilities of materials, to retrieve information irrespective of the heading in which the specification containing that information has been filed.

The United Kingdom classifies both European patent applications and PCT applications in the domestic system and this provides an additional entry point into the information contained in these important documents.

The classification system used by the United States Patent and Trademark Office is quite different from either the IPC or the United Kingdom system. It is the only official classification that can be used to search United States patents since, although they do apply the IPC, it is done by a concordance, and the results can be quite different from the IPC applied within Europe to equivalent specifications. The system breaks technology down into numbered classes which are further subdivided into numbered subclasses. The system is subjected to constant review and, at present, the classes of most interest to the polymer and plastics industry are numbered between 520 and 528.

Whilst these official classification systems provide a very useful search tool when using official collections of patent documents, their value is greatly enhanced when they can be combined with other search parameters in data bases produced by services such as WPI, *Chemical Abstracts*, the Claims files of IFI/Plenum Data Company and the APIPAT file produced by the American Petroleum Institute.

Sources of patent information

Official sources

In addition to the patent offices and specialist libraries which hold collections of patent documents and their associated indexes, there is one important data base supplier who bridges the gap between the Patent Offices and the secondary services. This is INPADOC, the International Patent Document Centre, situated in Vienna, who collate and publish the bibliographic information on patents from 53 national and regional patent issuing authorities. In general, the earliest information in the data base dates from 1968 but the starting date varies from country to country.

The information is made available on microfiche in the form of five quarterly and four weekly services. The four weekly services correspond to, and are cumulated into four of the quarterly service. The fifth quarterly service, the patent family service, is not available to public libraries but can be accessed online in Vienna or on various online data base vendors.

The services available provide (1) a patent number service which lists both applicant and inventor for each document and provides a useful method for tracing patent documents which are published in random order, (2) an applicants' data base which lists patents alphabetically by applicants' name, the patents for each applicant being listed by IPC, country of publication and type of document, (3) an inventors' data base constructed as for (2), and (4) a patent classification data base arranged by IPC which lists under each patent issuing authority those documents which have the same IPC. One disadvantage with the service is that the titles are reproduced as filed in the original language of the documents.

The fiches are cumulated each year and then again to cover the longer periods 1968–72, 1973–77, 1978–82 and 1983–86.

Secondary sources

The Derwent *World patents index* (WPI) is the major secondary service because it provides abstracts of the patent documents of 31 patent issuing authorities and deals with them in a systematic manner. The service also includes some literature which is related to patents, such as *Defensive publications and research disclosures*. The service defines as a basic patent the first patent document received disclosing a new invention. Subsequent patent documents from the same family disclosing the same invention in other countries are called equivalents and linked together by Derwent into families. Derwent produce a short 'alerting' abstract of each basic patent which concentrates on the subject matter of the claims and then a longer 'documentation' abstract which covers the disclosures of the whole specification in considerable detail. Derwent also give each patent an extended, informative title in English using controlled language. Coding systems have been developed, one originally designed for manual retrieval methods and a second, a punch card-based code designed for retrieval using punched card sorters. This mechanical method has been replaced by computerized retrieval but the code still has the same basic structure despite several revisions. Both forms of coding are available only to subscribers to the coded sections. Some of the Derwent hard copy services and microfiche services can be inspected in specialist libraries catering for patents.

Chemical Abstracts cover patent documents extensively but the abstracts tend to be less detailed than those of Derwent. However, *Chemical Abstracts* concentrate more on the body of the specification and the abstracts often provide information which is complementary to that in WPI. A major feature of the *Chemical Abstracts* service is the Registry Number system which is unique and provides an effective and efficient entry into the chemical literature including patents. Registry Numbers exist for many polymer and copolymer systems and the recent decision to cover separately block, graft and alternating polymers will increase the usefulness. The existence of open and patent literature in one data base is especially valuable when performing state-of-the-art and novelty searches.

The American Petroleum Institute provide two data bases, a literature file (APILIT) and a patents file (APIPAT). Whilst predominantly concerned with the petroleum industry, there is a considerable amount of information in these data bases which is of interest to the polymer and plastics industry. The abstracts now used in APIPAT are provided by Derwent, but the Institute applies its own detailed deep indexing and uses a system of linking to enhance retrieval.

United States patents are covered by a number of data bases but probably the best known and most valuable are the IFI/Plenum Data Company Claims files. There are several files of which the Uniterm files and the Comprehensive Database are especially useful. The Comprehensive Database has terms covering most monomers and a substantial number of specific homopolymers and copolymers. A sophisticated system of roles is helpful in increasing the relevance of the information retrieved.

Other patent data bases available online are the Japio file for Japanese patents, the new Chinese patent file, the French patent file, German patent files and the European Patent Office file. Some of these are of limited value because they contain mainly bibliographic data which is often available elsewhere with additional, more useful information.

A valuable file which is now available online is the EDOC file on Telesystemes Questel which is the special IPC file used by the searching section of the European patent office. This classification is somewhat different from the official IPC which appears on the front page of the patent documents and actually incorporates part of the old Netherlands Patent Office classification. The importance of this file is that it can be used to anticipate the prior art that will be found by the examiner and that the file is reclassified to reflect all changes in the key which overcomes the time-ranging which may be necessary when using the formal IPC. A corresponding key to this classification is provided in a second file, the ECLATX file, but the value of this is limited at present because it does not include the part written in Dutch.

There are several other specialist data bases available online which include patent documents but they will be of little interest to the majority of searchers. Many of the data bases mentioned are available on a number of online services and the service chosen will often depend on the data bases present, on the availability crossfile searching facilities and on the search language.

Searching for patent information

Information from patents can be obtained in one of three ways: the inquirer can (1) carry out the search manually using information provided by the Patent Offices and secondary services, (2) carry out the search online first then use manual searching if necessary, or (3) have the search carried out by an information specialist.

Name searches

These can be carried out manually using official name indexes provided by the patent offices of the countries of interest or by looking at the INPADOC or WPI microfiche. The Science Reference and Information Service (SRIS) of the British Library in High Holborn, London, has extensive holdings of the official indexes and the microfiche. Similarly, in the United Kingdom, the libraries which form the Patent Information Network (PIN) hold name indexes. Many European countries have similar regional libraries. *Chemical Abstracts* indexes also provide a useful means for name searching.

The use of hard copy indexes often makes it possible to pick up variations in the names being searched which is more difficult online. Online files of particular importance are WPI, *Chemical Abstracts* and INPADOC. Because of the variations in name that can occur, it is necessary to check the name index carefully using an 'expand' or like command. In addition to these three files which cover several countries, there are the files from the national patent offices and several versions of the United States Patent Office file.

Patent family searches

These can be carried out manually using the official name indexes mentioned above but this is not an easy task. Manual searches are best carried out using the INPADOC or WPI microfiche or, for earlier periods, the *Chemical Abstracts* patent concordance. However, today, the online files provide the most effective and most efficient way of tracing members of a patent family.

There are two files of value here, the first is WPI which covers most countries of interest. In this file, searches can be carried out using a variety of entries such as priority date, priority application number, local application number, company name or subject and details of the family are included in most printouts. The second major file is INPADOC. This can be accessed directly in Vienna in which case the access points are rather restricted, being priority application number or patent number, or on several of the online vendors. The WPI file will often include members of non-convention families, that is equivalent patents filed outside the priority year which are not linked together on INPADOC. However, it is possible to trace complex, multi-priority applications more easily on INPADOC. The INPADOC family service is rather expensive but it does cover 53 countries and it can supply legal status information which can make it cost effective when carrying out infringement searches.

Subject matter searches

Thorough subject matter searches through the patent literature are never easy and this is particularly true for the polymer and plastics area. One difficulty that arises is the rather loose terminology used within the industry. For example, what is meant by the phrase 'polymer and plastics'? The United Kingdom Patent Office definition of C3N and C3V indicates the extent of this problem. Most accept that thermoplastics are high molecular weight materials, in fact, macromolecular compounds. But are thermosetting resins macromolecular compounds? Patents concerned with epoxide resins appear in the macromolecular part of most classifications but those concerned with the manufacture or purification of the diglycidyl ether of bisphenol A (the archetypal epoxide resin), are often classified in the section dealing with low molecular weight organic chemicals.

If subject matter searches are to be carried out manually, the searcher must first decide which data bases are to be used. If official patent office files are appropriate, then the correct search term or terms must be selected from the IPC, the United Kingdom or the United States classification. In each case there are keyword or catchword indexes which will help in selection held in most libraries which stock patents.

In the UK, the staff at SRIS will give helpful guidance in this selection and, even more useful, the Classification Section of the Patent Office can be approached and will suggest a search strategy. Once the searcher has decided on the strategy, he can get hold of lists of specifications which have been assigned these terms or can go to a patent office which keeps a classified set of documents. The United States, German and Netherlands Patent Offices, for example, have such sets in their public search room.

In the UK, the searcher has two ways of working. The first is to go to the SRIS or another PIN library to search the books of abridgments and abstracts for those patent specifications which have been assigned the term or terms selected. There are indexes at the front of each volume which can be used for this purpose. Alternatively, he can order from the Patent Office file lists of all specifications (UK, EPC and PCT) which have been assigned the term or terms. This list can be used to search the abridgments and abstracts as above or to go through the specifications themselves.

The SRIS can supply lists of USA patents by class whilst lists of documents arranged by the internal European patent office classification are available from the Netherlands Patent Office. The INPADOC and WPI microfiche arranged by the IPC can also be used to prepare lists of specifications to be searched.

It is important to remember the philosophy behind official classifications when using them for private searching. Whilst many relevant patent specifications will have attracted the classification term that the searcher thinks appropriate, others will have been filed in a different part of the key because the examiner believes that reflects most closely the nub of the invention.

For example, EP 0098077 discloses a multistage polymerisation process for propylene using a catalyst prepared by reacting titanium tetrachloride with an organoaluminium compound. The IPC terms assigned are: C08F 04/64, which indicates a polymerisation catalyst containing titanium, zirconium, hafnium or compounds thereof and C08F 10/06, production of a homopolymer or copolymer from propene.

This illustrates the need to read the notes for firstly, whilst there is a term C08F 04/52 for polymerisation catalysts containing aluminium, this is not used because it ranks before C08F 04/64 and the last place rule applies. EP 0197310 (Fig. 5.1) attracted C08F 04/64 also despite the presence of the magnesium dihalide in the complex catalyst. Secondly, since the invention is concerned with the production of both propene homopolymers and copolymers it was assigned the classification C08F 10/06. Had it been concerned solely with the production of propene homopolymers, the term used would have been C08F 110/06 whilst an invention concerned only with propene copolymers would have attracted the term C08F 210/06. EP 0197310 also covers the preparation of olefin polymers but, since the disclosure is of a general nature, it attracted the general term C08F 10/00.

So, a search for information on the manufacture of propene polymers using the IPC would require a strategy which included all terms for specific types of propene polymer plus all the general terms for olefin polymers.

EP 0198588 is concerned with an electrically insulating composition based on polybutylene but the IPC term applied is H01B 03/22, insulating materials based on hydrocarbons. A searcher looking for general uses for polybutylene would not find this patent easily using just the official classification. Offices can view the disclosures differently amongst themselves and give quite different classification terms and it is always best to use at least two official systems or to employ one of the secondary abstracting services who index in greater depth.

For manual subject matter searches the traditional secondary source is *Chemical Abstracts* but the small print can be a disadvantage along with the limited depth of indexing. *Chemical Abstracts* is available in many patent libraries together with Derwent material which can be searched by the public.

Online searching usually proves the better way for most subject matter searches. Databases such as WPI, *Chemical Abstracts*, Claims and APIPAT usually provide either the IPC or the US classification together with many other data elements such as inventors, companies, and subject matter in the form of searchable index or key terms, title words and, often, abstracts. By combining terms and words, the output from a search is reduced and the relevance increased. WPI gives the IPC terms from all members of the patent family which can help to overcome the inconsistency between national offices. Additionally, inclusion of the prior art found by the examiner can aid extension of the search into areas not previously included in the strategy.

The CAS-Online file and the Telesystemes Questel DARC file

provide limited means to carry out structural searches in the polymer area. The structure of the monomer is used together with screens to carry out the search. This type of search will lead to a list of Registry Numbers which can be used to interrogate the normal *Chemical Abstract* files.

For those who are unable or unwilling to undertake searches themselves, there are several ways of obtaining information from patents.

Third party search services

Various patent offices will carry out searches for the public using their own internal search files, online services or both. The Australian, Austrian, Danish, Swedish and United Kingdom offices will all carry out technical and state-of-the-art searches. The United Kingdom office will also advise about patentability of an idea. Most offices will also provide a tailored SDI service in which specifications classified in specific parts of their system are sent to clients regularly.

The SRIS provides a search service which will search online files for clients whilst Derwent have bureaux in London and The United States Patent Office which will carry out searches through WPI. In addition to these, there are private searchers working in several patent offices and, in the Netherlands, Polyresearch Service BV, who, together with an increasing number of information brokers, will make patent searches for clients.

CHAPTER SIX

Standards — worldwide

J. S. WIDDOWSON

Introduction

Standards worldwide is a challenging but relevant theme since standards, technical regulations and the maintenance of quality are particularly important in the plastics sector of industry. Standards and regulations are complex documents with unique features necessitating the development of sophisticated information services to ensure their fullest exploitation. Frequently a number of sources may need to be consulted to ensure the quality or suitability of a product or service, ranging from Presidential Decrees, Richtlinien or Statutory Instruments for example, emerging from a large number and variety of international organizations.

Definitions

Standard

A standard is a technical specification or other document available to the public, drawn up with the co-operation and consensus or general approval of representative interested parties. Standards reflect their combined and consolidated data and experience and are intended to promote optimum community benefits. Approval is conferred by a body recognized at national, regional or international level (BS 4778 and International Organisation for Standards (ISO) Guide 2 1980).

Regulation

This is a binding document which contains legislative, regulatory or administrative rules and which is adopted and published by an authority legally vested with the necessary power.

Between these two classes of document a range of technical requirements has been established with varying titles and levels of enforcement, collectively defined below.

Technical requirement

This is any stipulation promulgated by an authoritative organization which affects the design, manufacture, use or application of a material, item of equipment or service and related systems providing proof of compliance.

The latter, relating to quality in the context of 'fitness for purpose', is becoming increasingly important to provide the user with proof that the product does indeed meet the standard. A variety of different quality assurance procedures are followed, from type testing to a full procedure involving inspection of the company, testing and repeat testing of randomly acquired samples leading to the issuing of a Kitemark or similar licence mark. Compliance with BS 5750 (Quality systems) or ISO 9000 (Quality management and quality assurance standards: guidelines for selection and use) provides another way in which quality can be assured by the introduction of a comprehensive quality management system which assesses everything from competence of management and employees through to the mechanism for tracing and rectifying defects in components which, when a company has complied, results in the certification of a registered firm or stockist.

So beneath the word 'standard', like an iceberg, is a mass of detailed work involving an array of expertise, all of which is dedicated to producing a product or service which is fit for its purpose and is most appropriate for the customers' needs.

The production of standards, the operation of testing and quality assurance (QA) services are simply facets of 'working for quality' with the provision of information as the mechanism by which most clients benefit.

The world's technical requirements

It might be supposed that an item of equipment or material, if safe in one country, would be equally safe in any other country, but this is not necessarily true. Plastic insulation for an electrical conductor, which works admirably in the temperate Mediterranean, might crack in the frozen north or become unstable in the tropics. Such climate-dependent variations will need different technical requirements, as specified in BS 6751:1986 (Protection of machine tools intended for use in extreme environmental conditions).

Cultural, religious and traditional requirements can also affect the sort of design which is acceptable to a particular country or political grouping. The Koran contains stipulations for the protection of the family and the location of holy places, which substantially affect the design of buildings. There have been occasions when design codes for protection against earthquakes, specially produced for a particular country, have not been adopted because of political factors.

It must be recognized, however, that there are also artificial 'technical barriers to trade' where the technical requirements have more to do with the protection of home industry than the elimination of any real hazard. This is the problem presented to the exporting manufacturer, to identify exactly what must be complied with to supply successfully a product to an overseas country. Technical barriers to trade are on the decrease owing to the activities at international level of GATT (General Agreement on Tariffs and Trade) and its standards code, and the provisions of Article 100 of the Treaty of Rome, with the European Community declaring that by 1992 all technical barriers to trade will be eliminated. But national preference, as often indicated in national standards, may still feature in a product which is successfully exploited in a foreign market.

The need for standards

Everyone needing protection from hazard, convenience of connection and unbiased technical advice benefits from the application of standards. Protection is necessary both from the ignorant or careless either of whom might introduce a dangerous feature for the sake of cheapness or because of lack of forethought or knowledge.

Standardization of design is essential for compatibility in, for example, components such as electrical plugs and sockets, car wheels and products such as microcomputer software.

Advice can be expensive, particularly when wrong. The 'individual' expert may be knowledgeable but may also be guessing! The system that produces standards eliminates unsafe advice by the need for a consensus of all expert committee members before the standard is made available to the public.

But the benefits do not end there, standardization aims to provide a means of communication amongst all concerned. It reduces the waste of effort traditionally associated with 're-inventing the wheel', it ensures that the consumer's interests are protected, and it improves the quality of life and the promotion of trade.

Standards in industry are particularly important in reducing costs by providing current data on which reliable designs can be based. Such data benefits everyone including the general management, who can assess its impact upon technological development and ensure its appropriate introduction within an environment defined by quality principles, productivity, and safety goals. The Design Department can exploit performance criteria, test methods, calculation methods, etc. to optimize design. Purchasing and Storage functions within the organization benefit from variety reduction by eliminating unnecessary duplication. The Production Department exploits the universal language of standards and the mechanism for controlling quality. Furthermore, the Marketing function can promote the quality of the company's products following the successful completion of relevant QA procedures and finally, the supply and distribution of products is assisted by standard commercial documentation, maintenance contracts and from the 'standard' nature of the product.

How do standards work?

Standards work because of the quality of expertise and the operation of sound techniques for developing the standard. BS 0: Part 2 'BSI and its committee procedures' discloses in detail the way in which standards are developed and committees are selected, staffed and operated.

Standards cover different categories of information such as glossaries, methods of all kinds, detailed specifications and codes of practice, guides and recommendations. A logical plan is essential to ensure that the terms are defined before they are used and the methods of test established and proved before QA procedures are instituted.

But, at the end of the day, standards work because of the contribution made by people. The British Standards Institution (BSI) as an example of a national standards body, can only co-ordinate the expertise made available by industry, government, consumers and all other areas of national competence.

An excellent primer is produced by the ISO (1983) *Standardisation and documentation — an introduction for documentalists and librarians*. This gives an historical perspective, amplified definitions and examples of physical documents from a selection of countries. Further data on the organisation of a standards information centre, the work of ISONET and descriptions of international and European standardization activities are also included.

Universal standards

The logic of having a single core set of standards for the world with deviations only based on climatic, geographical and cultural differences seems obvious when considering the benefits of standardization and variety reduction. The creation of ISO in 1947 may well have been stimulated by this aim.

However, the International Electrotechnical Commission (IEC) founded in 1906, produced its own electrical standards with the result that from the start, divisions for technical or historical reasons are evident. Moves in the 1980s have created structures which could lead to a common operation before long. For the time being however, we have a complicated and fractured world situation with many thousands of organizations which have produced over half a million documents.

There are, however, factors against which the likely significance of any series of technical requirements may be assessed to identify whether they are relevant to your needs.

Enforcement

Earlier definitions have indicated the category of 'voluntary consensus standards' whose adoption is voluntary unless they are cited by a legal authority or stipulated in a commercial contract.

On the other hand, the 'mandatory technical requirement' included in legislation is enforced by the legal structure of the country, through whatever system of primary and secondary legislation is in operation. In particular, governments tend to regulate against 'hazards'.

REGULATORY EXAMPLES

Occupational Safety and Health Acts — These cover conditions of employment and workers' safety by stipulating requirements such as machine guarding, the design of electrical installations, particularly in hazardous areas, such as paint spray booths, mines and underground railways. Specific 'Construction and Use' regulations for boilers, pressure vessels and the design of safety equipment for workers, such as hard hats and ear protectors, may also be included.

Building Regulations — These ensure the stability of buildings under earthquake, wind, snow-loading, etc. and identify systems for satisfactory access and fire escapes, emergency lighting equipment, as well as the permitted wiring systems, gas insulation and other facilities.

Food and Drug Acts — These cover processing, packaging, transport, storage and distribution of food or drug products, permitted or prohibited additives, such as preservatives and colouring matters and specific marking and labelling requirements.

Consumer Protection Acts — These are concerned with motor vehicles, domestic appliances and all perceived dangerous household items.

COMMERCIAL ENFORCEMENT

A further category of enforcement can be identified, that of commercial pressure. If, for instance, the national gas utility has its own standards and approval system and handles the majority of domestic appliance sales, then though that approval system is not mandatory, few manufacturers can afford not to comply. Similarly, where users of pressure vessel equipment or similar capital goods have joined together and produced a common specification, it is unlikely that many products will be sold which do not comply with it. Purchasing habits can also enforce requirements. If, for example, consumers expect to see a particular agency approval mark on an electrical product before buying, the impact on sales may force the manufacturer to comply.

SPECIFICATIONS

For convenience, documents produced by other standards-creating bodies are often called specifications. These can be produced by an array of differently motivated organisations. They are normally non-governmental and gain their significance by extent of use and status. Specification producing bodies may be testing organisations such as American Society for Testing and Materials (ASTM), Underwriters Laboratories (UL), etc. who need specifications in order to carry out their test work. Insurance companies, such as Lloyds, Det Norsk Veritas and the BG's produce specifications enforced on any company wishing to have their premises insured. Professional associations, such as American Society of Mechanical Engineers (ASME), Verein Deutscher Ingenieur (VDI), Verband Deutscher Elektrotechniker (VDE), the Institution for Electrical

Engineers (IEE) etc. can produce technical codes of practice which are binding on their members. User groups, consumer societies, research associations and many others can encourage compliance with the features they consider important to their group by the promulgation of specifications.

Geo-political factors

Most of the examples cited so far have been at national level, but in reality levels greater or smaller than this are equally significant.

INTERNATIONAL LEVEL

At an international level documents produced by the United Nations agencies, such as the Economic Commission for Europe, Food and Agriculture Organisation (FAO), Codex Elementarus, World Health Organisation (WHO), International Labour Organisation (ILO) etc., produce technical requirements that are wide-ranging in their application, but sometimes lack effective enforcement. Supra-national groups, with political status such as the European Economic Community (EEC), Middle East Gulf Co-operative Council and also, the United States of America, have the power to ensure enforcement and the regulations, therefore, are most likely to be observed.

USA state authorities for example, may take the Federal Authority to the Supreme Court to object to the introduction of a new Regulation. Voluntary collectives, such as Comité Européen de Normalisation (CEN), European Organisation for Electrotechnical Standardisation (CENELEC), African Regional Organisation for Standardisation (ARSO), Arab Organisation for Standardisation and Metrology (ASMO), Council for Mutual Economic Assistance (CMEA), Comisión Panamericana de Normas Técnicas (COPANT) etc., as voluntary groupings of standards bodies, can encourage co-operation, but normally only with national or political backing do these become other than voluntary documents.

SUB-NATIONAL LEVEL

Below the national level, division can be significant, as is evident in the provinces of Canada and Australia, the states of the USA and the Länder of Germany. In the UK, substantial technical differences exist between the building requirements in Scotland and those generally applicable in England. At state, county (and even town) level, therefore, variations in technical requirements can affect the product.

Application

Within a country it is feasible for technical requirement systems to be established which are intended to differentiate between the needs of different levels of user. In the Communist world, for example, three standards levels are recognized, with a fourth, Republic Standards, a legal sub-set of the state level — see *Standardisation and documentation, an introduction for documentalists and librarians* — these being:

State — applicable to the nation as a whole.

Industry — where sectors or branches are considered to require standards for their own particular category of industry.

Enterprise — where the individual standards required by a particular company are set before its products can be made.

Though less formal in outlook, the West recognizes some of these categories, particularly company standards which are normally only available from individual companies and which may be treated as confidential documents.

Scale of the problem

The number of organisations producing standards varies from country to country.

UNITED KINGDOM

The United Kingdom situation is straightforward due to BSI establishing a national engineering standards committee (the forerunner of the BSI) as long ago as 1901 to handle electrical and other sectors before diversification could develop. *BSI, the story of standards* by C. Douglas Woodward is a very readable history of these events.

WEST GERMANY

By comparison, West Germany (which many people think has only the Deutsches Institut für Normung (DIN) standards body) has over 180 organizations involved in the production of technical requirements — which explains the need for the government-funded, comprehensive technical requirements data base, DITR, which was needed as much to help German industry as those outside.

OTHER COUNTRIES

France has over 80 organisations and the USA many thousands, owing to its great size and also the tendency for industry to discourage the establishment of governmental standards activity by the creation of its own industry-wide organizations — leading to much diversification.

The future

The recognition of the harm caused by technical barriers to trade has provided motivation for substantial harmonisation of existing standards and prevention of the creation of new ones.

GATT, the UN agency General Agreement on Tariffs and Trade, through its GATT standards code, the creation of GATT enquiry points and mandatory circulation of drafts of any new technical barrier in advance of its implementation, has started to reduce the scale of the problem. Article 100 of the Treaty of Rome of the European Community has similarly provided a mechanism for simplification and harmonisation.

Successful industry is increasingly recognizing the benefit of single product standards and, where the standards world cannot move rapidly enough, they have occasionally carried out their own work, for example, the so-called 'High Sierra' group in providing elements of standardisation for CD-ROM production.

The logic of harmonisation is evident but the national political wish to protect its own work force from extensive competition from cheaper, more effective or higher quality products from other areas is also obvious. However, overall the trend is towards reduction.

Finding Relevant Technical Requirements

The solution to identifying and obtaining technical requirements relevant to your problems depends largely on your occupation and on the resources you have available. A professional librarian or information scientist for example working in a large, well-funded information unit will have different problems from an engineer on site in the north of Scotland.

However, whoever you are, remember when working with technical requirements you must strive to comply with the three Cs:- Completeness, Currency and Comprehensibility.

Completeness

Ignorance of the law is no excuse and all relevant technical requirements *must* be retrieved or the answer could be misleading or even dangerous.

Currency

Technical requirements change with time. The collection of documents which are enforceable or relevant to you today may well be different from that identified only weeks ago, so you or the source you use must be capable of rapidly obtaining updated information on all relevant standards, regulations, approval practices, etc.

Comprehensibility

Having obtained the technical requirements you need to analyse and work to them. If you can't read or make sense of them you still have a problem. Authoritative translations of foreign documents are critical and you need an agency that produces 'technically equivalent statements' not just linguistic translations or your organisation may misunderstand and re-design the product in a way that is not acceptable to the importing country or client.

Expert advice on the precise meaning and method of implementation of the requirement can also be critical and services like THE in the United Kingdom and Norex in France exist for just that purpose.

On the information side the best handbook on standards information is the ISO/UNESCO publication *Access to Standards Information — how to enquire or be informed about standards and technical regulations available world-wide* (1986), produced by Eric Sutter of AFNOR, with particularly good coverage in France.

Options for technical requirements information

CATALOGUES

The majority of standards-producing bodies have an annual catalogue and you might think that organisations believing in standardization would have catalogues with a common presentation but unfortunately they all differ.

The BSI Catalogue (previously called the Year Book) is a particularly good example, with a full listing by document number complete with abstract and history, cross-reference tables to European and international standards and a logical, comprehensive subject index produced by information specialists. While it is published only in English and is by far the most exhaustive National Standards catalogue available, it does not cover Technical Requirements available in the UK.

The West German standards organisation, DIN, has produced a

comprehensive catalogue (in two bulky volumes) covering all Technical Requirements of Länder as well as at Federal and international level, in German, with some in English, following substantial funding from the German government. While lacking abstracts the catalogue provides comprehensive information, although the subject structure based on modified UDC is not easy to use. However, sub-titles in English and a keyword index are a useful feature.

The ISO Catalogue, designed for standards makers is structured under Technical Committees but nevertheless has a broad subject index as well.

Association Français de Normalisation (AFNOR), the French standards body, has a catalogue structured by a classification of its own with a separate English version. Outside of the major organisations however, the problems become more difficult both in presentation and language. *Access to standards information* has a full list of what is produced.

Some improvement is under way through ISONET whose members, even at the basic Type 1 membership level, are required to produce a standards list in English or French based on titles with at least an annual update.

With specifications, the producing organization decides how best to list its products, usually in a sales format.

Legal documents lists often cover the total field of legislation, so identifying relevant technical requirements can present problems.

CATALOGUE UPDATE SYSTEMS

Currency is critical and the annual catalogue is often out-of-date before it is published. As a result, most major organisations offer an update service of some sort, often as a commercial service.

In BSI the *Sales bulletin* provides a bimonthly cumulative listing now identical in presentation and subject index to the annual catalogue, thus ensuring that current, complete and comprehensible information is available. For monthly information *BSI News* includes a total listing of all documents that will be available by the end of the stated month. It includes details of reprints, special announcements, documents proposed for confirmation, reviewed and confirmed, standards proposed for 'declaration of obsolescence', declared obsolescent for withdrawal, withdrawn or on which new work has started.

International new work, details on drafts for public comment, draft common names for pesticides, draft international standards, international publications and European scale documents are also detailed. It even includes amendments to the *Buyers guide* and organizations in the UK who obtained type certification against Canadian standards and is probably the most comprehensive tool of its kind in the world.

COLLATING DOCUMENTS

Given the scale of the problem of identifying related Technical Requirements some publications offering special assistance are available.

CORRESPONDENCE TABLES

The European Standardisation Committee (CEN) produces an annual directory *National application of European standards*. This publication attempts to list equivalent national standards for the 16 members of CEN together with bibliographic data.

North Atlantic Treaty Organisation (NATO) also produces documents which cross-relate different standards, but these tend to have a limited circulation.

INDEXES

The KWIC index of international standards produced by ISO identifies all existing international standards in a given field but is not updated during the year.

Industry standards locator index, produced by IHS (for subscribers only) provides a simplistic index to USA, ISO, IEC and some European standards organisations' publications. Use of titles only can cause loss of relevant documents. For example, building code does not take you far into the technical content of the document — earthquake design, snow and wind load etc. Other USA documents exist but none provide comprehensive access to all Technical Requirements.

The EEC ICONE project produces a *Comparative index of standards* which aims to cross-reference standards produced by the national members of the EEC where related by international or European reference.

Selected bibliographies are produced by ISO, DIN and other larger standards bodies which are useful provided your interests match the selected subject and you remember to update the information from other sources.

TECHNICAL PROGRAMMES

As part of the standards-making activity programmes have to be produced and these documents can provide valuable advice on future standards and work-in-hand. GATT under its 'Standards Code' also requires notification of any draft which could constitute a technical barrier to trade.

COMMERCIAL INFORMATION SERVICES

Manual systems are complicated, labour intensive and require substantial time, tenacity and money to prepare and to ensure compliance with the three Cs. For the end-user a better solution may be to pay for the information sought.

BSI INFORMATION SERVICES

BSI has evolved a number of services including:
PLUS (Private List Updating Services) — an updating supply service for collections of standards. Following a list of all British, international and (selected) foreign standards (DIN English, API, and UL), BSI automatically supplies each month the relevant revisions, amendments, notices of withdrawn or obsolete standards and a basic new products list and order form. Shortly, a link with BSI's bibliographic data base STANDARDLINE is planned to give an individual complete selective new publication service.
BSI Sectional Lists — a basic subject listing biased to the functions of technical committees. Sectional lists are soon to be supplemented by more comprehensive information products extracted from STANDARDLINE.
OSUS — the Overseas Standards Update Service is similar to PLUS but provides world-wide monthly information on change. The actual documents can be ordered from the BSI Sales Department. It covers all significant Technical Requirements received in the BSI library with an initial structuring charge to validate, correct and set up your original list and an annual service charge, both fees depending on the number of documents.
BSI World-wide Standards Information Service — for a subject approach to current information the BSI Library exploits its substantial updating activity against a charge relating to the number of subject entries and countries.

BSI therefore produces a significant number of commercial information services from internal information systems, but most of these are available only to BSI members.

Similar services and publications are available from many of the major standards bodies with standing-order systems being particularly popular, often against national classified structures, but online services and developments from them are the most fruitful future for better information, albeit with a higher price tag.

The information industry

The information industry also has a part to play in standards information. Information Handling Services and Technical Indexes have a major role in providing microform systems for many of the European and American voluntary standards. Though expensive, the convenience and guaranteed 28-day update provide the ultimate referral collection but user reaction to microforms is not always enthusiastic.

For special subject areas a variety of paper products exists, such as:- Stahlschluessel — Key to Steel — a tri-lingual listing of German steels matched against 20 other countries in an attempt to enable equivalence to be identified.
World Standards Mutual Speedy Finder — compares Japanese with USA, UK, Germany, France and USSR steel characteristics.

Terminological information is available from INFOTERM, the International Information Centre for Terminology, operated in Vienna in conjunction with UNESCO, affiliated to the Osterreichisches Normungsinstitut (ON).

Commercial publishers, such as Pirola, Direction des Journaux Officielles, and others produce compendia collating legislation and regulations together on particular subjects. The Bureau of National Affairs in the USA provides a particularly comprehensive collection of commercial services giving access to secondary legislation in occupational safety and health, consumer products and environmental areas, with the US government as a major client.

Mechanized services

On-line, bibliographic data bases could have been designed to solve the Technical Requirements information problems. With the fundamental need for currency and completeness, the ability to change data on a central computer and make it available instantly to any user throughout the world eliminates at a stroke many of the problems associated with paper services. However, the currency and completeness of the particular data base used may not be adequate. For example, Orbit Standards Search data base was updated at only three month intervals.

Some 20 years ago, before the technology was well established the International Standards Organisation formed an executive committee, INFCO, the Committee on Information. While handling a variety of conventional information matters on exchange of translation information, classification systems etc., it became clear that what was really required was a world system for standards information covering all Technical Requirements. The plan was that each national member organisation would take responsibility for the identification and description of standards and technical regulations operational in its geographical area (with ISO for international standards) and cover them on an internationally 'standard' information system using a common, multi-lingual thesaurus.

ISONET data bases

Following extensive work over many years ISONET (ISO Information Network) has been established with a common instruction manual and two officially recognized thesauri, the AFNOR, ITT Thesaurus and the BSI ROOT Thesaurus.

There are now eight public access ISONET data bases covering the UK, Canada, Denmark, Germany, France, Sweden, Finland, Norway and more are under development. The Japanese system is undergoing field trials.

ADVANTAGES OF ISONET DATA BASES

Technical Requirements are many and varied and to miss one is critical, so when the national standards body creates an ISONET data base, quality is assured. DITR is an excellent example in which every conceivable Technical Requirement has been included in the data base. Furthermore, the information staff have access to the committee secretaries, chairmen and members who understand the contents of the standard and can provide the necessary technical support.

THE ISONET STANDARD SYSTEM

To gain officially graded ISONET membership, compliance with the mandatory manual is essential, but since the system is decentralised, the national need for complete data ensures that a comprehensive file is established. The ISONET Manual (first edition 1985) should be used if you have a need to describe your own organization's standards or similar documents, since the analysis carried out was thorough, and the information description of a Technical Requirement is comprehensive. However, some confusion exists because there are two official ISONET thesauri. A French core document was subsequently developed and improved by BSI who introduced a subject display with facet indicators. The recent book *Thesaurus construction* by Jean Aitchison and Alan Gilchrist describes the thesaurus system. The BSI document was eventually published as the BSI ROOT

Thesaurus in English with a French manuscript translation available. Subsequent support from the relevant national Standards bodies has resulted in the published versions available in German (DIN), Czech (CSN), Japanese (JSA) and Chinese (CCIS), and an Arabic version is nearing completion. A recent ISONET survey identified 13 ISONET members as having adopted the ROOT Thesaurus.

The French TIT (ITT) thesaurus is available in French and Spanish with an English translation, with Polish and Serbo-Croat versions available in due course. The 1987 survey identified 5 countries using TIT with one contemplating a change to ROOT. It is hoped that as a result of a new EEC project (see section, The Future, later) a common correspondence table if not a total merging will be produced. However, the availability of two multilingual controlled vocabularies is a substantial advantage over most information systems.

Constant feedback has been established to improve and correct each system and the plans to merge or introduce cross-file searching will ensure precise equivalence.

However, even with substantial internal and external work loads, activities such as translations and maintaining significant holdings in a library etc., a major commitment to standards information exists in most standards bodies. Overall, there is a national supply responsibility to provide easy access to copies of all national standards, with supply or re-direction to other sources for specifications on technical regulations.

Consultancy or advice services are also often available, related to the Information Centre so that a full service is available to resolve all problems.

Continuous development of the mechanized information systems ensures their progressive improvement and also the identification of further needs, and new systems to make the data even more readily available. Enlightened self-interest is evident with the need to identify and sell standards but there is a genuine belief within the ISONET world that people are better off if they can easily find the standards they need.

Comprehensive information on ISONET is obtainable from the ISO Central Secretariat in Geneva. Copies of the ISONET Constitution and ISONET Directory (1987) contain a full listing of all the organizations who support the ISONET system (well over 70). The contact address is Case Postale 56, CH-1211 Geneve, Switzerland.

The ISONET *Manual*, first edition 1985, is also invaluable to both users and creators of ISONET data bases because it leads to the definitions of the data elements that are available.

The ISONET *Guide* (currently under revision) is a background document explaining how various aspects of the services work, but is still of interest to those developing their own standards information services.

The ISONET *Communiqué*, as a regular newsletter, is received by all ISONET members.

Other on-line information systems

Other sources for Technical Requirements information also exist, though some may not meet the high standards set by the ISONET data bases.

For a basic listing BSI Information Department sells, through its library, a *Brief guide to standards data bases*, identifying and detailing the ISONET data bases and four non-ISONET American files.

A document issued by the Sveriges Standardiseringskommission (SIS) in autumn 1986, at the 1985 ISONET Workshop, went further including *COMPENDEX (computerised engineering index)* and *Easinet* as an example of a simpler mechanism to access some of the available standards.

The ISO document *Access to standards information* has good coverage of data bases but other systems can also be of help: LEXIS — LEXIS is now the principal legal online service, but is usually restricted to the legal professionals. However, if you have a friendly contact in the legal world it can be useful, with comprehensive coverage of English law, including Statutory Instruments, where many of the regulations are covered. Growing files covering Irish, USA and French libraries make it progressively more valuable and Scottish law is also being added.

These are not comprehensive and they may not go back as far as you need in order to see the regulations which affect your product. BSI has also had some difficulty in tracing certain British Standards which we understood were cited in legislation.

Other online services can help such as the US file for the Federal Register (*Federal register abstracts*, Dialog file 136), Bureau of National Affairs files are valuable, covering occupational safety and health, consumer product liability and many other sectors of the complicated regulatory scene in the USA.

The EEC is covered on many of the scientific files, such as INSPEC (electrical engineering) PIRA (packaging and paper), METADEX (metals) etc. All cover aspects of standards and regulations, but caution is needed as they may not be complete, current or comprehensive enough for your purposes.

For other sources of online information you need to look at *UK online search services*, published by Aslib or similar world directories.

Quality assurance (QA) information

Standards are of value without an external proof of compliance, but if you have it, you have proof positive that the product is made to standard.

A definition of quality, QA and related terminology may be found in BS 4778: Part 1: 1987 (equivalent to ISO 8402:1986) for international terms and BS 4778: Part 2 (as modified in AMD 5641, 29 May 1987) for national terms.

For more basic information perhaps the catalogues produced by organizations involved in QA matters could be useful. For example, BSI Quality Assurance, Certification and Assessment Department produces a collection of leaflets with titles like *The way to capture new markets* which briefly describe BS 5750: Quality Systems, the quality standard, assessment and registration both established and through QASAR, product certification (Kitemark, Safety Mark etc.), special certification schemes (BS 9000 electronic components, stockists, validation and manufacturers' data etc.), *Now that you are certified*, and *Quality in action* a quarterly journal on BSI QA matters.

Lloyds, Yarsley and other QA organisations have similar publications.

PRODUCT CERTIFICATION/REGISTERED FIRMS' DATA

Users of products or firms need to know if a target company is certified for the products they wish to buy or has been registered as a firm with the required QA and industrial and technical skills.

The normal source for this information is a Buyers Guide. The 86/87 BSI publication contains a subject index against the appropriate British Standard number, a product certification list showing the standard, an informative abstract of the certification features and a listing of the companies who have been certified against that standard. Registers of firms of assessed capability for manufacturing industry sectors, service industries, individual firms and registered stockists, PASCAL compiler validations are all listed against the quality assessment schedule number, full names and addresses of companies.

A similar listing for the Swiss Association for Quality Assessment is also included with a final index of firms against appropriate QAS, BS or Swiss number. This comprehensive document is invaluable, but is current only to 1 June 1986. Perusal of all subsequent *BSI news* is required to update this information.

Other organizations with an active assessment and certification role produce their own publications such as the British Board of Agrément and the Property Services Agency. The latter produces a useful publication *Selling building products for PSA work* which lists relevant standards, quality assurance schemes, registered firms schemes and BASEC (British Authority Service for Electrical Cables) and CARES scheme (Certification Authority for Reinforcing Steels) as well as the BSI requirements for BDA certified products — useful but since it ends at May 1985 it requires updating.

For more general data the Department of Trade and Industry's *Register of firms of quality assessed capability* lists all known firms assessed by organizations such as the MoD, TCB, BSI etc., but its currency and future is in doubt.

Similar publications to these are produced by some foreign approval or relevant government agencies but they can be difficult to acquire and maintain current.

COMPUTER-BASED SERVICES

An online service would be ideal, but to date the Promolog proposal by the European Community produced by AFNOR which was planned to be a data bank of certified or approved products has appeared only as a European listing of organizations with a book index to identify them.

CODUS, a valuable specialist data bank, marketed by Technical Indexes Ltd., exploits the BS 9000 and European Committee for the Co-ordination of Electrotechnical Standards — Electronic Components Committee (CECC) electronic components systems and permits the selection of appropriate references to approved components against technical data of the component required.

Standards for plastics and polymers

For the first time in this chapter, some three-quarters of the way through, specific reference is being made to plastics and polymers and you may wonder why.

Principally, the reason is because the world of technical requirements tends to work from hazard to product and, therefore the critical requirements with which you need to comply may involve you in a total search of all available documents. For example, a pressure vessel regulation may identify that the produce must be safe and of a good design and provide criteria from which you must decide if your plastic component can meet those requirements.

If you contemplate the construction of processing plant in the UK or abroad the general occupational safety and health requirements will be applied to your project and again you must decide whether you can comply. Consumer requirements, again, may be defined in terms of flammability, toxicity, etc., and the responsibility rests with you to confirm the suitability of the product against the Regulation.

Therefore, all the finding aids identified above could be relevant to your particular search. However, for specific standards questions, the procedures outlined below are suggested.

UK data on standards

BRITISH STANDARDS INFORMATION

In the United Kingdom, BSI has the major responsibility for the development of standards in all fields, which include plastics and polymers.

For the manual searcher the Catalogue is the first logical place to look. Using the new quality subject book index, the subject heading plastics has a 'see also' reference to polymers, thermosetting polymers, ultra-high molecular weight polymers. Browsing through words beginning with 'poly' you will find numerous entries under polyamide fibres, polyesters, polyethylene, polyolefin fibres, polystyrene, etc. Reference to the *Sales bulletin* provides cumulative bimonthly updates.

The Sectional List covering plastics and rubber (SL 12) is obviously useful but the latest issue of May 1986 will clearly not contain reference to subsequent documents. An update search on STANDARDLINE showed that of the 709 documents, 110 had been changed, with four revisions, 21 amendments and 85 where the status had changed — so the need to search for the latest information is essential. Structured in a way that reflects Technical Committees activities, SL 12 covers interests in aircraft material, cellular materials, coated fabrics, flooring, hose, laminated sheets and tubes, mechanical rubber goods, methods of test (9 close-packed pages), moulding, extrusion materials and resins, pipes and fittings, raw rubber, rubber and plastic footwear, rubber goods for hospitals, sheet plastics, standards for electrical use and an unclassified catch-all section which may or may not match your interests.

However, selection of a particular standard such as BS 4962 — *Specification for plastic pipes for use as light duty sub-soil drains* — provides information on its history in the foreword (reference to previous MAFF field drainage experimental requirements) relationships with international work, and a cross-reference list to related standards at the back.

For more current information a trawl through the relevant issues of *BSI news* will be required.

The BSI world-wide standards information (WSI service) can help by automatically providing information on whatever subject or country of interest based on updated information received in the BSI Library. For example, the issue of October 1986 detailed changes of eight standards in six countries under the heading Plastics and a further 16 standards from eight countries (a different eight) for plastics as a material. More detailed information is available under specialist headings such as silicon compounds, seals, coated textiles etc., and, of course, the documents are available for loan to members.

STANDARDLINE

The alternative to the paper chase is using the BSI STANDARDLINE data base. This will provide guaranteed complete and current information, updated in advance of any paper publication and with the full sophistication of 39 data elements available for search. A comprehensive introduction pack is available from BSI, Milton Keynes.

A STANDARDLINE search for all standards referring to plastics or polymers (using PLASTIC* or POLY* truncated) found 889 standards. Six months later in June 1987, 52 new or revised standards in the fields of the polymers or plastics were retrieved. They covered every conceivable field from flushing cisterns, polystyrene, moulding and extrusion materials, children's cots, medicine measures, plastic reinforcement for fabrics, road and rail tanker hoses, dental elastic impression materials, rotating electrical machines, non-metallic spur gears, safety harnesses, trade cutlery, tile flooring, aerospace jointing, wall tiles, conveyer belts and even dustbin lids — so the necessity to look for updated information is paramount.

If you don't have direct access to STANDARDLINE then seek a professional intermediary. BSI's Enquiry Service offers help but will charge for more complicated searches requiring use of the sophisticated data base.

Other sources such as your professional institution, local

technical branch of your public library or industrial information services such as HERTIS will also help but again usually for a charge.

But as you are aware the problem doesn't stop with standards — regulations and other documents could be relevant.

LEGISLATION

On legislation, a relevant data base is POLIS, which contains information on British parliamentary questions, proceedings, and papers as well as legislation and official publications. EEC and other relevant information from foreign official publications are also included.

For a further demonstration, a search for plastic toys was made which found that the concept 'novelties' was needed since the 1980 and 1985 Regulations are entitled Novelties (Safety) Regulations and its amendment. The Food Imitation (Safety) Regulations and Toy Water Snakes (Safety) Order were retrieved as well as a number of Parliamentary questions. The Toys (Safety) Regulations 1974 did not appear in POLIS because the legislation data only started in November 1982, so such limitations must be borne in mind.

The result showed that the use of the term plastics was not helpful but the follow-up using the term toys and games was more successful retrieving five relevant items — a useful result. The original publications however must still be acquired. LEXIS (a USA legal data base) will also be useful if the comments made earlier are taken into consideration.

Reverting to manual methods an enquiry of the UK GATT Regulations Enquiry Point operated by the Department of Trade and Industry could identify British requirements at the barrier-to-trade level, or again, an approach to the BSI Enquiry Service, RAPRA Technology Ltd., or other specialist trade associations such as the British Box and Packaging Association, British Plastics Windows Group, is often useful. Such organisations may be found listed in the CBO *Directory of British associations*.

Foreign plastics and polymers standards information

The world is your oyster, any of the world's half-million standards and regulations could contain relevant data. The primary source in the UK is the BSI Information Department at Milton Keynes and for members many of the services are free or available at a discount. Through the THE Service (non-members will be charged) access to comprehensive consultant assistance may be obtained.

GERMAN DATA

In Germany you often have the world's most complicated situation, but with DITR the national data base made available by the German Standards body DIN, the solution is comparatively simple.

A search of DITR both from the UK, by staff at BSI, and from DIN staff produced the same results. This equivalence of search has much to do with the common thesaurus, DIN having produced a German language version of the BSI ROOT Thesaurus. The same precise search mechanism and relationship with the original indexes is therefore available in both English and German.

The search for plastic toys found one specific DIN standard (DIN 53160, June 1974) concerning the testing of coloured toys for resistance of saliva and perspiration.

Broadening the search to toys referenced in titles and in the classification system used by DIN produced 37 documents, not only DINs but VG, DIN/VDE, LM, EG, ZHI, RAL standards as well as items of legislation, but inspection is necessary to assess the significance of the plastics element.

An alternative to the online search is to use the DIN Catalogue which is produced from the DITR data base. Results using the Catalogue classified section found plastics in 6760 and toys in 7250, but which also includes saddlery, travel goods and camping equipment. By inspection DIN 3160 duly appeared in the toy section but not in the plastics classified section as the title did not include the word.

For other plastics questions sections 6770, Testing of plastics; 6780, Testing of mechanical properties of plastics; 6790, Testing of material defects and of physical and chemical properties of plastics; 6800, Plastic products, semi-finished products; 6810, Testing of plastic products; 6820, Auxiliary materials and additives for plastics; 6830, Plant and machinery for plastics industries and 6832, Cellular materials, could have been searched.

US DATA

The USA is a valuable source of information, though many of their information systems concentrate only on the American market.

As an example, *Information sources*, the annual directory of the Information Industry Association, is a useful source of relevant information services. PLASPEC is identified as an engineering and marketing data bank and claims to provide substantial data from company specifications to enable appropriate plastics to be selected but there is limited reference to National Standards.

Armed with bibliographic listing you now need to obtain copies of the listed documents and as with Germany this requires a multiple approach. Again, the BSI Library will have most, if not all, of the documents and many available in translations, at least in the form of title and list of contents. Complete quality translations are also available from THE.

MULTINATIONAL AND INTERNATIONAL INFORMATION

In the EEC the CELEX Database can help, but again it is restricted to European-scale legislation and does not cover the national related documents which in many instances could affect your enquirer.

Internationally, the various UN agencies have information services and the CIAS service from the ILO or those operated by WHO, FAO etc., are valuable, but they tend to rely on free supply of information from other agencies and therefore currency can be in doubt.

The International Standards Organisation in Geneva has a role within ISONET to help on all international publications, so either their service should be consulted or their KWIC Index of International Standards used.

World standard information of plastics and polymers creates a problem because of the multiplicity of sources needed to be consulted. A do-it-yourself exercise may be personally satisfying but the time involved often makes it more cost-effective to use the professional intermediary services available from BSI or others in their specialist areas.

CONSULTANT/INFORMATION SERVICES

Definitely, for commercial enquiries involving the export of products, the use of the THE service should be seen as mandatory, as failure to comply with the relevant foreign technical requirements could result in refusal to import your product, however good it may otherwise be.

The equivalent THE services operated from France (NOREX), New Zealand or the special service from the Underwriters Laboratories are alternatives but they tend to specialize in the provision of information to nationals of their own countries, so THE is normally the answer for the British enquirer.

The future

What's wrong with the present?

The size of the section dealing with tools to find standards and regulations show the sort of thing that is wrong with the present — far too many sources, all different, all requiring great effort to identify relevant information. In general they are:-

COMPLICATED

Even the best of them require special training and in-depth knowledge of the information system to make them work. Knowledge to enable application of the data will always be necessary but that is the professional skill of the scientist or engineer concerned which must be acquired. In some senses the systems have been designed for the professional intermediary and not the man with the problem.

IRRELEVANT

Bibliographic data identifying further papers which must be sought, extracted, read, understood and often discarded as not being relevant, is inappropriate. More data banks, full-text files and specialist agencies, capable of resolving real problems, should be developed.

MIS-DIRECTED

The end-user needs the information but it is often filtered through a Librarian or an Information Officer who may not 'add value'. Where the project is new and outside the principal experience of the end-user then obviously the professional intermediary is the right person to provide information. However, where the data is required to refine existing knowledge or extend what is already known, then it is of little value to require the expert to explain to the intermediary his problems in detail and for the intermediary to debrief the expert. Direct end-user systems — designed by the professional information person but intended for the manager, contracts engineer, marketing staff, designer, research scientist, standards engineer, i.e., the professional end-user — must be developed.

Scenario for the future

At the technical level an unsafe product should never be designed.

Increasingly, standards and Technical Requirements will be built into computer-aided design systems so that basic safety and compatibility features will always be included.

For world markets a system of international standards will link the national requirements with climatic or cultural variants of the universal standard and legislation will cite these standards so that world trade can be facilitated.

Information systems themselves will be available as simply and conveniently as the telephone and the pocket notebook. The system logic will be human, communications — quite possibly — by voice and the system will have expertise built-in to be able to handle 'fuzzy' logic and enable the enquirer to move from internal data bases to any external data bases and data banks. The terminal will enable the user to record the results, redirect to colleagues, or have the answers assessed by expert systems.

New products — now

But let us think more immediately, what will be available tomorrow? The CD-ROM (compact disk, read-only memory) is already with us and is being developed with the end-user in mind. Bought as a package with reader and micro, all relevant data and software systems are part of the purchased or leased package. No longer will the end-user be dissuaded from searching by unfamiliar systems and high expenditure. CD-ROM will encourage extensive use so that the full extent of the ROM library can be exploited to resolve search problems. Updating may well come online and translation in English and all major languages, if not universal translation, will be available through multilingual thesauri.

Bibliographic data bases will progressively give way to full text. The best (ISONET) current refined bibliographic indexes will be the essential mechanism to retrieve data from the full text systems, given the impossibility of controlling the native language of the author or editor.

END-USER SYSTEMS AND SERVICES

Increasingly, information will be generated with the end-users' retrieval systems in mind. Consistent vocabulary and the use of standard data will be progressively introduced to enhance these systems.

Technical requirements will be harmonized, at international and regional levels. GATT and regional political groupings such as the Gulf Co-operative Council in the Middle East, point the way forward. 1992 is the deadline for the EEC to have eliminated all internal barriers.

The future will therefore permit the end-user to be more creative, engaged solely in the development of cost-effective, fit-for-purpose products. The use of more sophisticated QA procedures requiring compliance with improved, relevant standards, utilizing efficient test procedures will ensure that the product on service has the quality you want, hopefully at a price you can afford.

ISONET will probably still have a part to play. A major investigation is currently under way to assess what the user wants from standards information and to tailor services further towards this end by working in conjunction with the information industry and other active parties. The standards world, of course, also has an active role to play in developing the standards to which information systems and products are themselves produced. If you have special needs or interests, contact your national ISONET member.

In conclusion, people in the plastics and polymers industry have the same problems as everybody else with Technical Requirements. They must look hard and long at all aspects which might affect the fitness of their products for both national and foreign service. Great persistence is required to achieve the three Cs but this, coupled with contact with expert sources of information, notably at BSI, will help your clients produce or use the best quality equipment or services.

CHAPTER SEVEN

Trade literature, theses, conferences

J. STUBBINGTON

Within the considerable number of information sources available on polymers and plastics, the trade literature sector seems to be more neglected than most. The main reason for this unfortunate circumstance is that frequently the organization and control of the material are poor because it is targetted on users whose need of traditional sources of technical information is perhaps lower than that of the research and development scientist.

The literature is copious, it has no standardized format or regular publication pattern and is thus a daunting prospect for the information professional. More so than for the non-professional in this area, the task of tracing, acquiring and keeping the collection current requires a dedicated commitment to its potential value. Although often produced in a variety of colours it truly belongs to the elusive material popularly known as 'grey literature'.

The emphasis of this chapter is to highlight some of the problems encountered in dealing with trade literature and to suggest ways of minimizing these and to draw attention to the key sources and services for tracing and acquiring this material.

It would be as well to define 'trade literature' at the outset, as there are many company publications which will not strictly come within the scope of this discussion. Broadly speaking it is any publication issued by or on behalf of an organization in order to promote its image or products. It is important because it may contain original information which is not published elsewhere. It is also designed with a care and attention to detail reflecting the large sums which are spent to further one of its prime purposes of product marketing and image promotion.

Increasingly, the technical comprehensiveness and depth of the information supplied make it imperative that it should not be ignored but actively exploited. The better systems do attempt to make the information as easily accessible as possible, for example through the indexing and classification systems developed by RAPRA Technology Ltd. Other large collections, notably that of the Science Reference Information Service (SRIS) are essentially repositories only.

Formats

Part of the problem with trade literature is the wide variety of formats in which it is produced ranging from single data sheets to multivolume sets. Special forms include customer handbooks, formularies, service manuals, user guides, catalogues and directories. Some are so substantial that they gain the status of recognized reference works or handbooks.

Apart from the obvious need to assist the marketing of products by eye-catching sales appeal frequently there is a wealth of technical detail given to encourage existing users to find new applications or markets. This data may never appear elsewhere and although the charge of selectivity of data may be made nevertheless it provides a source which should not be overlooked. The combination of lavish use of colour, both in line drawings and photographs enhances the visual impact of the presentation, thus communicating its message in ways which journal articles and reports seldom manage to achieve.

Bibliographical control

As trade literature is not issued through normal commercial publishing channels, bibliographical control is minimal. Regular announcements of new or updated trade literature however may be found in the trade press often accompanied by enquiry service cards which may be used to select the desired items.

It requires a good system of journal scanning management however to be certain that all relevant items are obtained, particularly for items from overseas. The larger companies themselves need control over the substantial amounts of literature they produce and elaborate check lists and indexes may be produced. ICI for example issues a regularly updated *Technical Service Note (No. G101)* which lists all trade literature produced on grades of polymers and plastics materials.

Other companies implement coding systems for their products which are useful for the identification and subsequent organization of the collection but the onus of acquisition and updating, with some notable exceptions, is firmly in the hands of users. A good source of trade literature which has been systematically acquired over the years, while still not totally comprehensive is the collection available through the *RAPRA Abstracts* as a printed monthly journal or as an online data base offered by Pergamon-Orbit-Infoline.

The SRIS too, has an extensive collection of product literature arranged alphabetically by company obtained from over 20,000 mainly British companies, including a collection of earlier material from about 1830 to 1940. One of the most comprehensive collections of product literature and catalogues, chiefly on engineering, which are regularly and systematically updated is produced on microfilm by Technical Indexes Ltd. This information is classified, indexed in detail and backed up with an enquiry service. The Engineering Components and Materials Index leads to a coded chart on which a detailed classification of the products is to be found and which in turn leads to the microfilm cassette and frame numbers. A microfilm reader/printer may be leased from Technical Indexes as part of the service.

An excellent overview of the whole field of product literature and its exploitation appears in *Finding and using production information* (Wall, 1986). This work discusses sources for inter alia plastics, paint, rubbers, elastomers and engineering materials, e.g. composites, in great depth.

House journals

These are serial publications produced by the larger industrial companies, business houses and public sector organizations usually for image-promoting purposes for both employees and clients but nevertheless they often contain useful information on the structure, products and activities of the organisation. In the USA they are known as house organs.

The style and content of the house journal ranges from prestige-seeking journals emphasizing the scientific and technical achievements of the organization aimed at specialists in the field, to magazines containing general and non-technical articles as well as reports on company product developments.

Regularly appearing product catalogues may also have some of the characteristics of the house journal proper. Irregular publication is often one of these and although issue and volume numbers, and indeed an ISSN may be allocated, they can be difficult to trace. Some help may be obtained however from the following guides:

Benn's Press Directory, published annually covers 780 UK titles;

British Association of Industrial Editors — Membership Directory covers 800 UK titles;
The Standard Periodical Directory, published (bi)annually, covers 5000 USA and Canadian titles;
National Research Bureau — The Working Press of the Nation covers 3600 USA and Canadian titles.

On the whole, house journals are a difficult medium from which to extract information as their bibliographic characterization leaves much to be desired. They are often published irregularly, with inadequate numbering and dating, title changes, and widely different circulation policies to suit the perceived needs, usually market and image-building, of the producer.

Systematic collection and exploitation is therefore beyond the resources of many technical libraries to cope with, so tracking down and using some of the more important titles published by companies in the polymers and plastics industry mentioned below will require great persistence.

The following selection lists some of the better known house journals all of which contain company and product news.
Avon News (IH Publications Ltd.) is a bimonthly newspaper from the Avon group which is a good source of company and product news.
BFG Citizen (B. F. Goodrich) monthly newspaper focuses on company and product information.
BHRA News (BHRA Fluid Engineering) is a quarterly newsletter giving company and product news supplemented by an enquiry card service.

A good example of the genre is the *BIBRA Bulletin* (British Industrial Biological Research Association) circulated to members as a 50-page monthly journal containing BIBRA news items, an editorial, research news and results, world news, meetings and a useful subject index.
BPF News (British Plastics Federation) (ISSN 0142/7148) includes feature articles on current industry and market topics, product information, announcements and reports of meetings, exhibitions and personnel movements.
Battelle Today (Battelle Institute Ltd.) is a controlled circulation, monthly subscription journal for company news, special reports and short technical articles.
Bayer Reports (Bayer AG) is the biannual English equivalent of *Bayer Berichte*.
CVP Resinotes (Cray Valley Products) is a useful source of technical data, properties and applications for new products, and research and development activities.
Drager Review (Drager Safety), monthly, is the official journal of

Dragerwerk AG containing news on recent legislation as well as company and product news.

Du Pont Magazine, European Edition (Du Pont de Nemours International) is produced quarterly in English, French, German and Italian and contains company business results and product news.

DuPont Engineering Design (Du Pont de Nemours International) contains brief technical items, applications and product news.

ECHO (Du Pont de Nemours International) is an irregularly issued review of Du Pont wire and cable products.

Elastomers Notebook (Du Pont de Nemours International) is a well established (1938–) journal covering products based on Du Pont raw materials and providing a materials selection guide. Customer test results may be included and also longer technical reviews. An enquiry service is also available.

Fulmer Update (Fulmer Research Institute Ltd.) contains product and applications news items and is particularly useful for lists of contracts.

Goodyear Clan (Goodyear Tire and Rubber Co. (GB) Ltd.) is basically a staff newsletter but nevertheless contains company matters which can be of interest.

ICI Engineering Plastics (Imperial Chemical Industries plc.), produced quarterly in newspaper format, with summaries in French, German, Italian, Spanish and Swedish, contains brief technical items on products and applications.

ICI The Roundel (Imperial Chemical Industries plc.), is basically a business review covering ICI interests worldwide appearing five times a year.

ICI Polyurethanes Newsletter (Imperial Chemical Industries plc.) contains news of technical developments and uses of polyurethanes chemicals and is supplemented with market and product reviews.

Materials News International (Dow Corning Inc.) Available in English, German or French it is a valuable review of new silicone products and applications, including design and process techniques current in industry, science and medicine.

Merchandising Vision (British Cellophane Ltd.) contains product, applications and BCL company news.

Norsk Hydro (Norsk Hydro) Published quarterly in several language editions each issue focuses on a theme e.g. safety in the industry and the policy and technical developments of Norsk Hydro in this area.

PIRA News (PIRA) The newsletters are produced by the various divisions of PIRA to disseminate news of seminars, publications and people.

Plastics Today (Imperial Chemical Industries plc.) Subtitled the business magazine for the plastics industry this quarterly, available in English, French and German covers a wide range of product and application developments.
RAPRA News (RAPRA Technology Ltd.) is published quarterly and available free on request, carries news of current research and development carried out by RAPRA, meetings and publications.
Resin Review (Rohm and Haas) is a valuable source of information on products and research results.
Rohm and Haas Reporter (Rohm and Haas) is published quarterly and is a good source of product and application news.
SAMPE Journal (Society for the Advancement of Materials and Process Engineering) published bimonthly and available on subscription contains technical, contributed papers and news of conferences and publications together with a buyers' guide and an enquiry card service.
SAMPE Quarterly (Society for the Advancement of Materials and Process Engineering) Similar in style to the *Journal*, the *Quarterly* includes high quality contributed technical papers and a detailed index.
SATRA Bulletin (Shoe and Allied Trades Research Association) Issued monthly to members only, the *Bulletin* includes details of new processes, products, applications and news of research projects with accompanying technical data.
SEATAG Bulletin (South East Asia Trade Advisory Group) This quarterly bulletin is available free on request and carries information on industrial budgets, economic news and reviews of industrial developments. Government advice to exporters is also reported.
Shirley Institute Bulletin (Shirley Institute) Available to members only and published irregularly the bulletin covers Institute and membership news, with technical details of research programmes.
Spirax Topics (Spirax Sarco Ltd.) This is brief in the extreme, but occasionally reports technical matters.
Steetley News (Steetley Ltd.) This is a typical company newsletter in newspaper format reporting staff news, company progress and useful product data.
Vetrotex Fiberworks (Vetrotex UK Ltd.) Published in a glossy, image-promoting format the journal usefully covers a wide range of company news, details of commercial, product and applications developments.

Conclusions

House journals are one of many types of trade literature, but with often inadequate bibliographical data to help manage their acquisition and control, they may fail to reach the wider audience capable of making use of the often valuable commercial and technical information they contain.

Commercial trade literature services however, offer options such as consolidated catalogues of trade literature obtained from contributing companies, with a standardized format. The resulting compendia may be bound into somewhat unmanageable volumes. A good example of the genre is the *Thomcat* section of the familiar *American Thomas Register*. Other regularly updated trade catalogue services provide access to the collections through detailed indexes and classification schemes and are available as loose-leaf and microform files, e.g. the Technical Indexes collection of catalogues.

The ideal solution to the problems of obtaining access to all the relevant trade literature has yet to be devised. Requests for payment to include the literature in a catalogue immediately inhibits the willingness to contribute, as much by the wealthier as the less wealthy companies, even though the benefits in terms of marketing exposure of the product are fairly obvious.

In-house collections are valuable only if based on careful selection and appropriate dissemination, storage and indexing of the material. Standards for the bibliographical control of this somewhat ephemeral material have yet to be convincingly achieved. The commercially available systems come closest to the ideal and offer the greatest help in overcoming the many difficulties of obtaining and managing the trade literature.

Theses

The history, nature and purpose of theses have been well covered in *Theses and dissertations as information sources*, Donald Davinson, 1977 (Clive Bingley Ltd, London and Linnet Press, Connecticut). Valuable chapters discuss the bibliographic control of theses and access to them. For the UK the only comprehensive listing since 1950 has been the *ASLIB Index to Theses*, currently published (twice yearly) in classified order with an informative abstract. Broad classification headings of particular interest include *Polymers, polymerisation* — D1h; *Polymer technology* — L7e. For 1988 it is expected that over 120 abstracts of theses will be published under these headings. As with all sources within the polymers and plastics subject area however, useful information is also likely to be found under e.g. Organic Chemistry, Chemical

Engineering, Bioengineering, Powder Technology, Solid State Physics.

Dissertation Abstracts (University Microfilm International), covers theses in chemistry, engineering, and physics produced by the great majority of American and increasingly, Canadian and overseas universities, colleges and other institutions. This publication is also available as an online data base on *DIALOG* growing at the rate of 3500 records annually and giving access to nearly one million theses published since 1861.

British Reports, Translations and Theses (BRTT), monthly, is a 'bibliography of material falling within the category of 'grey literature', with cumulated annual indexes' (Russon, 1988). British Government organizations, local government, university, and learned institution reports and translations are listed and doctoral theses accepted after 1970. With the exception of translations all the material is mounted on the *SIGLE* data base (System for Information on Grey Literature in Europe) on Blaise.

Conferences, symposia and meetings

Attendance at these events is often a highly agreeable means of making informal contact with acknowledged experts and acquiring information well in advance of publication. The broad 'state-of-the-art' surveys on specific themes are probably difficult to obtain by any other means. The scope and scale of meetings may range from prestigious international gatherings lasting for several days to small sectional meetings of professional or technological societies. The larger meetings may cover a number of themes, The American Chemical Society Division of Polymer Chemistry's latest symposia for example, was (co)sponsored by eight organizations for the different topics covered. The events themselves are usually well publicized in advance in journal diaries of forthcoming events, by individual mail shots e.g. from RAPRA Technology Ltd., the Plastics and Rubber Institute (PRI), British Plastics Federation (BPF), Engineering Index and in the *ASLIB Forthcoming International Scientific and Technical Conferences*. Events announced in this quarterly journal are indexed by subject, location and organization.

Papers presented at the meetings may be pre-printed, if not in full, as abstracts, or published later as full proceedings, including ideally, discussion arising from the presentation.

The major conferences at least, receive full abstract coverage appropriately in, for example, *Engineering Index*, which covers over 1800 conferences each year, *RAPRA Abstracts, Chemical*

Abstracts, covering the American Chemical Society Divisions e.g. Polymer Chemistry, Industrial and Engineering Chemistry and Organic Coatings and Plastics Chemistry Symposia. Proceedings of the PRI conferences are published regularly, with previous issues being made available at lower prices. The British Library Document Supply Centre has an extensive collection of conference proceedings, fully indexed by keyterms taken from the title and earlier editions available as cumulated indexes on microfiche. This collection is also available on Blaise as the *Conference Proceedings Index*. A full discussion of earlier conference literature may be found in Yescombe (1976).

References

Wall, R.A. (ed.) (1986) *Finding and using product information — from trade catalogues to consumer systems* (Gower).

Yescombe, E. (1976) *Plastics and rubber: world sources of information* (Applied Science).

CHAPTER EIGHT

Online data bases

R. T. ADKINS

Development of computer-based typesetting for the production of printed abstracting and indexing services led to a gradual realisation that the machine-readable version could be searched at a remote location through the increasing availability of the international telecommunications network. Searching at its most basic comprises matching the index terms in a query against the term used to index the item on the data base. In its simplest form this process was carried out in batch mode by software capable of handling a number of queries during the serial passage of the abstracts on tape through the computer. Production of inverted indexes on disk (where the item references are stored under the appropriate index term) led to development of powerful online command languages permitting interactive searching, a process of modifying the search strategy in response to the output obtained until the desired result is achieved.

Over the past 15–20 years the provision of online retrieval services has developed into a highly competitive industry. There are currently thousands of data bases (over 3000 are listed in the *Directory of Online Databases* (Cuadra Associates/Elsevier), 1987, which is also available online through Orbit. Similarly, the DIALOG online file, *Database of Databases*, containing information on some 2800 data bases, is published as *Computer Readable Databases: A Directory and Sourcebook: Science, Technology and Medicine*, M.E. Williams, (Information Retrieval Research Laboratory). *Accessible Databases*, eds. D.A. Tookey and G. Mortimer (Spicer and Pegler Associates), 1988, covers 799 data bases worldwide, particularly business data bases, including bibliographic, full-text, numeric and transactional. This rapid growth of the online industry has added further complexity to the problems faced by users in their search for information.

But in spite of the disadvantages of high cost, the great diversity of file structure, command language and telecommunications protocols, the benefits of availability, rapidity, flexibility, comprehensiveness and currency, are undeniable. Furthermore, many of the larger data bases are loaded on to more than one host, which helps to limit the need to learn a large number of command languages. Rather than attract the end-user of the information, searching on the whole has tended to remain within the province of the specialist intermediary who may be employed by the organisation or whose expertise is made available through publicly available services. RAPRA Technology Ltd., the Science Reference Information Service and Derwent Publications Ltd., for example, offer fee-based searches on demand by skilled intermediaries.

The *Marquis Who's Who Directory of Online Professionals* (Marquis Who's Who Inc., Chicago), gives details of e.g. over 300 online specialists in the field of Materials Science, their expertise with hosts, data bases and equipment, career history and publications.

Complementing the growth of the online services, a burgeoning literature has developed from which descriptions of the features, scope and limitations of data bases and techniques for searching them may be obtained. Useful periodicals include *Online Review* (Learned Information), *Online Notes* (ASLIB), *UKOLUG Newsletter* (Institute of Information Scientists Online User Group) and *Information World Review* (Learned Information), monthly. Frequent seminars and exhibitions, organized by ASLIB, The Institute of Information Scientists, the Science Reference Information Service and the Library Association, cover the needs of the tyro and experienced user. The annual International Online Meeting and exhibition held in December by Learned Information Ltd., is the major event of its kind in Europe, supplemented by product reviews and published proceedings.

More comprehensive bibliographies may be found in *Online Bibliographic Databases — a Directory and Sourcebook*, J.L. Hall (ASLIB), 1987, and in the *Manual of Online Search Strategies*, eds. C.J. Armstrong and J.A. Large, (Gower), 1988.

The UKOLUG Guide to Online Commands, A. Arthur (Institute of Information Scientists), 1987, is a handy guide to the main commands of the major online hosts, while for those who are starting up, a good beginners guide is *Going Online*, G. Turpie (ASLIB), 1987. The *Online Business Sourcebook*, eds. A. Foster and G. Smith (Headland Press), covers a wide range of sources of financial and company information.

The major data bases in the plastics and polymer field are in Table 8.1. Reference has also been made to the major online data bases in other chapters to which the reader is referred, notably *Chemical Abstracts, RAPRA Abstracts* (Ch. 3), *WPI, CLAIMS, INPADOC* (Ch. 5), *Standardline* (Ch. 6), business data bases (Ch. 13), *World Textile Abstracts* (Ch. 14), *World Surface Coatings Abstracts* (Ch. 16), *KKF Kunststoff Kautschuk Fasern* (Ch. 18), Japanese data bases, (Ch. 21).

Some of the data bases are concerned solely with plastics and polymers while others are technologically more broadly based, tending to cover the whole of the chemical and engineering industries for example. Other more specialised data bases contain valuable information on the applications and business aspects of plastics materials. In short, possibly relevant information may be widely scattered throughout many data bases. Use of features such as the Dialindex File on DIALOG may give a better indication of the most appropriate data bases to search. From this file the number of postings for each indexing term used in data bases placed in broad subject categories may be determined, with sometimes surprising results.

Many data base producers publish detailed user guides, copies of thesauri and classification systems used for controlled indexing of the information, often with information of the number of postings to the index term or heading.

RAPRA Technology Ltd. have taken a further step in exploiting their data base by publishing the results of searches carried out on broad topics including blowing agents, and flame retardant agents for polymer foams, asbestos substitutes, PET bottles, UHMW polyethylene. The bibliographies are indexed by title, author, company and trade name.

CD-ROM (Compact Disk-Read Only Memory) data bases

It is early days yet for the full impact to be assessed of the steadily increasing numbers of CD-ROM data bases available on the position of the established online retrieval systems. Compared with the thousands of data bases available online less than 200 CD-ROM products are being marketed or planned for release in 1988 judging from the entries in the *CD-ROM Directory* (TFPL — Task Force Pro Libra), 1988. This is a thoroughly useful and up-to-date compendium of CD-ROM products. The brief introductory overview by editor Kate Churchill covers the present state-of-the-art well and offers some useful insights into the future development of the industry in relation to conventional online systems. Further items of interest include a bibliography of directories, textbooks and market research reports, relevant journals and conference reports.

The huge capacities of CD-ROM disks are well suited to the storage of works whose printed equivalents occupy many volumes and linear feet of shelf space. Although the data bases are large, software has been developed which allows retrieval at very high data rates (150 Kbps), giving access to the desired information within a few seconds.

Of particular interest are the *Encyclopaedia of polymer science and technology*, ed. H. Mark (John Wiley & Sons), 1987, in progress, the first five volumes being available on disk. Other encyclopaedias include the *Kirk-Othmer encyclopaedia of chemical technology*, (John Wiley & Sons), with all 25 printed volumes, index and supplements being on disk, the *McGraw-Hill science and technical reference set* corresponding to the 15 printed volumes of the *Concise encyclopaedia of science and technology* and the McGraw-Hill *Dictionary of scientific and technical terms* (McGraw-Hill Book Co.).

Other major scientific, technical and business data bases available include the *Science Citation Index* (Institute for Scientific Information), the *Applied Science & Technology Index*, the *General Science Index* and the *Business Periodicals Index*, (H.W. Wilson Co.). Many others are projected for release in the near future. The *UKOLUG Newsletter* has started a regular feature 'CD-ROMs Corner' to monitor what will be the undoubtedly rapid advances in this sector of the information industry.

Property and other data banks

Selection of materials for specific applications is as much an art as a science but the development of computer selection aids has eased the problems of choosing the right materials for the job.

Plascams 220 (Plastics Computer-aided Materials Selector) is a knowledge-based plastics materials selector system produced by RAPRA Technology Ltd. Values have been assigned to the properties of over 350 materials by a team of independent experts. Searches may be carried out on over 75 different properties by elimination procedures or by weighting the search to optimize the results in line with the values assigned to the data by the evaluating team. Materials include reinforced, foamed, filled or lubricated thermoplastics, filled or reinforced thermosets, and thermoplastic elastomers are also included. Searchable categories include

TABLE 8.1 Online bibliographic data bases

Data base	Producer	Host	Size Date range	Comments
APILIT APIPAT	American Petroleum Institute	Orbit	418000 182000 1964–	Literature data base. Covers Derwent CPI patents including Plasdoc.
CA Search	CAS	Datastar Dialog ESA/IRS Orbit	>15 m 1967	File structures differ with host and may be split by year. Worldwide literature/patents coverage.
CAS Online	CAS	STN		STN is a joint CAS/FIZ/JICST service. Full abstracts from 1970. Chemical (sub)structure searching available.
Chemical Business Newsbase	Royal Soc. Chemistry	DataStar Dialog Orbit	29000 1984	Bibliographic and factual data are searchable. Good coverage of Europe.
Chemical Industry Notes	CAS	DataStar Dialog Orbit	0.5 m 1974	Worldwide (esp. USA) coverage of chemical and plastics production.
CHINAPATS	Chinese Patents Office	Orbit	6300 1985	English language abstracts.
CLAIMS	IFI/Plenum	Dialog Orbit	1.7 m 1950	US Patents. Detailed chemical subject indexes.
CMAI Petrochem Market Reports	Chemical Market	IP Sharp Assoc.		Economic statistics and forecasts for chemicals and plastics industry.

Cordura	Cordura	Cordura	1977–	USA only — covers adhesives, elastomers, moulding and extrusion resins, foams. Commercial names and sources.
Compendex	Engineering Information Inc.	BRS DataStar Dialog ESA/IRS Orbit STN	2.2 m 1970–	Plastics & polymers engineering & technological literature. Now merged with Engineering meetings file.
DRI Databases	McGraw-Hill	Data Resources	30 m	Over 125 historical & forecast data bases
East European Monitor	DataStar	Business Intl., Vienna	10,000 1984	Business information on petrochemicals, plastics, fibres.
Engineered Materials Abstracts	ASM & Inst.Metals	DataStar Dialog Orbit	2900 198	Covers ceramics, polymers, for structural and non-structural uses; composites.
European Chem.News	ECN	DataStar	1984	Full-text of ECN online. Includes petrochemicals, plastics, rubber, fibres. File is updated the week before the journal is published.
IDC-Polymer Speicher	IDC		150000 1976	Plastics and polymer patents. Batch service only.
INPADOC	Intnl.Patent Doc. Cntre, Vienna	Dialog Orbit	13 m 1968–	Patent documentation from 52 patent offices.
INSPEC	IEE	BRS DataStar Dialog ESA/IRS Orbit	3.0 m 1967	Physics, electrical engineering, electronics, computers & control

Data base	Producer	Host	Size Date range	Comments
JOIS	various	JOIS		see Ch. 21
Kirk-Othmer	Wiley	BRS DataStar Dialog	25000 1984	Full text of the 3rd Edn. of the Encyclopedia of Chemical Technology. Covers plastics elastomers, fibres etc.
KKF	Deutsches Kunststoffe Institut	STN	1973–	Plastics, Rubber and Fibres see Ch. 18.
Materials Business File	Metal Information	Dialog DataStar ESA/IRS	0.25 m 1985–	Produced jointly by the Am. Soc. Metals and the Inst. of Metals. Covers engineered materials, polymers, plastics composites etc.
PASCAL	CNRS	ESA/IRS Questel		Multilingual science & technology data base.
PIRA	PIRA	Orbit	140000 1975–	Paper, printing, packaging. Document delivery service.
PLASPEC	Plastics Technology	ADP	current	Engineering, design & processing characteristics of plastics.
Plastic Chemicals	SAGE Data Inc	SAGE	various 1947–	Economic/technocommercial statistics on plastics.
Plastiserv	SPI	DB Systems	various	Product guide to plastics in USA.
POLYCAS	CAS	Questel	1985	Plastics & polymer, trade names & nomenclature from CAS Registry files.
PTS Promt	Predicasts	BRS Dialog DataStar	1.50 m 1972–	Markets & technology of i.a plastics, rubber, fibres, packaging worldwide esp. USA.

RAPRA	Rapra Technology Ltd.	Orbit	0.25 m 1972–	incl. Adhesives subfile. Core data base for science, technology and business information needs of the plastics and rubber industries.
RAPRA Tradenames	RAPRA Technology	Orbit	2500 1976	Materials, machinery, additives, etc.
SCISEARCH	ISI	BRS DataStar Dialog DIMDI Orbit	7.0 m 1974–	Citation searching possible.
Soviet Science & Technology	IFI/Plenum	Dialog	90000 1975–	
Standardline	British Standards Institution	Orbit	10,000 current documents in force	Standards, Codes of Practice, Drafts for comment, covering materials, engineering, Building, electronics, safety, testing etc.
				see Ch. 6.
World Textiles	Shirley Inst.	Dialog Orbit	155000 1970	see Ch. 14.
WPI	Derwent	Dialog Orbit	2.0 m 1966–	Section A (Plasdoc) covers plastics and rubber patents with comprehensive indexes.
WSCA	Paint Research Assocn.	Orbit	114000 1976–	see Ch. 16.

m = million.

general and electrical, mechanical, chemical and radiation resistance properties, production and post-processing methods.

Matus, available online from Engineering Information Company Ltd. covers properties of approximately 7000 engineering materials (plastics, metals, ceramics, composites and some natural products) taken from mainly UK suppliers' catalogues.

Peritus (Matsel Systems Ltd.) is further knowledge-based system for material selection, containing property data on plastics and metals.

PETS (GE Plastics Ltd.) is another example of the growing number of expert knowledge-based systems, capable in this instance of providing the user with appropriate corrective actions to problems which may be encountered in the moulding of GE Plastics products, e.g. silver streaks in polycarbonate resins or dark streaks/black spots in polyphenylene oxide resins.

The *CAPS* data base (Polydata Ltd.) covers 60 major manufacturers of thermoplastics, converting all data to comparable units on about 5000 grades with 100 data per grade, updated continuously and supplied to the user on PC-compatible disk. The search program comprises a user-friendly, menu-driven system and allows graphs and bar charts to be produced as output to the search.

PRAS, the *Plastics and Rubber Advisory Service* jointly run by the British Plastics Federation and the British Rubber Manufacturers Association offers an enquiry service based on their computer data base of information on about 800 companies and their product data. Set up in 1982 the service is free to members, non-members paying an annual fee for enquiries covering the full range of materials, products, equipment and services. The largest sector of the data base is concerned with packaging. Two recently announced computerized data base systems (1) providing technical data for the designer or fabricator in engineering thermoplastics are the BASF *Werkstoff Informations System* and *CATS (Customer Aided Technical Service)*, a materials comparison program produced by Du Pont.

References

Plastics and Rubber Weekly, 20 Feb 1988, p.28

PART II

CHAPTER NINE

Polymer structures and nomenclature

R. G. LINFORD

Introduction

Polymer nomenclature has been, and remains, a problem area. This is well illustrated by an example recently described by Vincent (1987). Poly(ethylene imine) is the molecule formed by polymerizing the imine CH_3-$CH{=}NH$ to give $(CH_2CH_2NH)_n$. Polyaziridine is the polymer formed by the ring-opening polymerization of

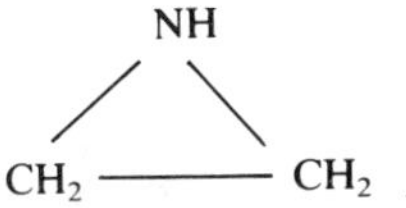

$$\begin{array}{c} NH \\ / \quad \backslash \\ CH_2 — CH_2 \end{array}$$

and this polymer is also $(CH_2CH_2NH)_n$! It is also possible for the same name to describe more than one polymeric species and confusion inevitably results (Loening, 1981). Fox (1968a) points out that poly(ethylene oxide) is frequently written in a way that omits the parentheses; the name can then refer to $HO(CH_2CH_2O)_nH$, or

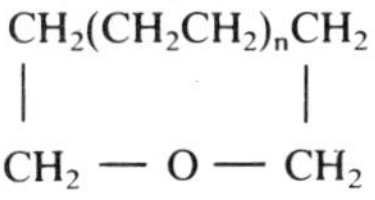

$$\begin{array}{ccc} CH_2(CH_2CH_2)_nCH_2 & & \\ | \qquad\qquad | & \text{or} & H(CH_2CH_2)_n \diagdown \\ CH_2 — O — CH_2 & & \qquad O \\ & & H(CH_2CH_2)_n \diagup \end{array}$$

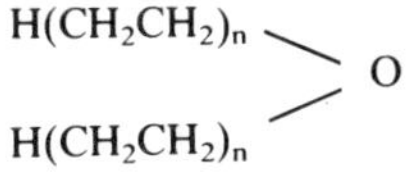

It is not unique to the polymer field to have the same entity described by several names, and the same name serving for several entities. Linford (1973) has reviewed the problem caused in the solid surface thermodynamics field by a similar confusion between names and parameters. In that case, as with polymer nomenclature, meaning and communication are both sacrificed for the sake of conformity to historical practice, a point also emphasized by Fox (1968a).

This brief survey of the difficulties for the information searcher that arise from problems of polymer nomenclature, is structured along the following lines. By way of introduction, types of polymer, polymer use and polymer information are considered and a history of nomenclature is sketched. The principles and practice of naming polymers are outlined. A selection of the difficulties within the search services available to the polymer user is described and finally the present day status and future requirements are reviewed.

Types of polymer

A specialist user within the polymer field has a clear idea of what the word 'polymer' encompasses; it includes the class of polymeric substances that he or she works on together with chosen examples from adjacent classes. For the non-specialist occasional user, such as a physical chemist wishing to extend a model by considering data on macromolecular analogues of his system of interest, or for the information scientist, the term 'polymer' is conspicuously broader. Examples of categories within the field include:

KEY TECHNOLOGICAL THERMOPLASTICS

In 1975, 86% of the world production of thermoplastics originated from four polymers: polyethylene, 40%; poly(vinyl chloride), 23%; polystyrene, 12%; polypropylene, 11%. Less than 2% of the thousands of new polymers synthesized each year are commercialized (Olabisi, 1982).

SPECIALITY HOMOPOLYMERS

These are molecules characterized by the repetition of a single chemical structural unit, whose relative molar mass or molecular weight, M_r, (Jenkins, 1984) is some multiple of that of the unit (Fox 1968a).

STEREOREGULAR POLYMERS

For certain types of polymer, identity of empirical chemical formula but difference of structure gives rise to clear distinguishable materials (Fox, 1968a; Glasse and Linford, 1987).

COPOLYMERS

These contain more than one type of structural unit, arranged randomly or with the units in alternating sequence, or in blocks, or perhaps formed by chain grafting into combs or networks (Fox, 1968a; Glasse and Linford, 1987). Some usages (e.g. Donaruma, 1986) restrict copolymer to two types of monomer unit, using number-specific terms such as terpolymer, tetrapolymer (or for IUPAC (1974a) usage, quaterpolymer), and so on for multi-monomer polymers.

POLYBLENDS

These are mixtures of homopolymers or copolymers and the same term will be used here to include the admixture of such additives as pigments, fillers, plasticizers, reinforcing fibres for polymer composites, and coupling agents (Olabisi, 1982).

POLYMERIC FIBRES

An important example is polyester (Davis and Hill, 1982).

POLYELECTROLYTES

These are macromolecules bearing a large number of ionizable groups and like uncharged polymers they may be linear, branched or cross-linked. They are used as separation membranes (Anderson and Morawitz, 1982; Smith, 1987).

POLYMER LIQUID CRYSTALS

Aramids, e.g. poly(*p*-phenyleneterephthalamide), and thermo*tropic polyesters, e.g. poly(p*-hydroxybenzoic acid-*co-p,p'*biphenol-*co*-terephthalic acid) exhibit liquid crystal behaviour, the former in solution and the latter in the melt. Solution of natural and synthetic polypeptides, and a wide range of nitrogen containing polymers such as polyazomethines and polyisocyanides also exhibit liquid crystal behaviour in solution (Kwolek et al., 1987).

POLYMER ELECTROLYTES

These are solutions of ionic salts in mechanically solid polymers such as poly(ethylene oxide); they behave as ionic conductors and electronic insulators and are exciting considerable present interest as potential components in batteries, sensors and electrochromic displays (Vincent, 1987; Owen, 1987).

CONDUCTING POLYMERS

These inherently electronically conducting materials such as polyacetylene, polyaniline and polypyrrole, are seen as vitally important in military, security and energy-conservation applications (MacDiarmid and Maxfield, 1987).

PROTEINS

These are biochemical polyelectrolytes (Anderson and Morawitz, 1982). All enzymes are proteins.

POLYMER LANGMUIR-BLODGETT FILMS

These are model analogues of lipid bilayers used to mimic vesicle behaviour and to act as ultra-thin gas permeation membranes (Albrecht et al., 1985).

INORGANIC POLYMERS

Silica, quartz and a range of other materials are macromolecular in nature. In additon, a range of non-carbon-containing polymers exist, such as $(SN)_x$ which is called polythiazyl or poly(sulphur nitride).

It can clearly be seen from the above, non-exhaustive, list that the diversity and complexity of polymer types adds to the task of producing a sensible and systematic nomenclature which will enable information to be retrieved from the literature. Sometimes the view is taken that polymers are distinguished from other macromolecules by the presence of a readily discernible repeat unit, the amounts of which are sufficient to provide a set of properties that do not vary markedly with the addition or removal of a few units (IUPAC, 1974a). On this definition many naturally occurring macromolecules such as proteins and polysaccharides would be excluded but others such as cellulose would be included (Donaruma, 1986). This definition would tend to include oligomers, i.e. substances with a small number of repeat units, within the polymer classification.

Type of information

Depending on the use to which the information is to be put, chemical information about precursor monomer and/or repeat unit may suffice. For stereoregular polymers and for co-polymers, various levels of structural information may also be vital. For many applications, relative molar mass (M_r) information is essential as short-chain length versions of the polymer may be liquid or waxy whereas longer-chain material may be solid. Occasionally the name is changed to aid classification; for example, poly(ethylene glycol) is poly(ethylene oxide) of low M_r.

Morphological information such as the degree of crystallinity may also be pertinent. Low density polyethylene, LDPE, has substantially different mechanical properties from the more crystalline, high density, HDPE, form of the same repeating unit. The effect on performance parameters such as conductivity, of the presence or absence of crystalline material within the amorphous matrix of polymer electrolytes of different stoichiometries and at different temperatures has been discussed by Linford, Neat and co-workers (1985, 1986), and present nomenclature is not sufficiently refined to take easy account of this.

Who needs polymer information?

The orientation of interest of the user dictates the type of information that they need and the nature of the information service to which they can turn for help. The polymer field is perhaps the most important example of a multi- and inter-disciplinary study and is penalised by the subject-based conceptual thinking, historically possessed by academics and passed on to some major information providers. For example, Chemical Abstracts Service, the world's largest bibliographic and substance registration data base, works on the premise that chemists are predominantly synthesisers or analysers and undeniably caters well for such needs. Those such as physical chemists, materials scientists or polymer technologists, who wish to study, use or modify polymeric materials, have to broaden their search strategies in order not to suffer from unexpected consequences of CAS policy, such as the abstracting only of selected articles from certain issues of particular journals.

Polymer-related information is both provided and used by a surprisingly wide cross-section of the technical public. This includes scientists, such as biochemists, electrochemists, physicists and materials scientists; technologists in such fields as building materials, plastics and rubber technology, membrane separation processes and energy conservation; artists and architects pursuing novel texture; environmentalists interested in biodegradability and waste processing and consumers desiring well packaged foods and other commodities.

The nomenclature needs of each type of user are quite different.

Scientists require a systematic approach to remove potential ambiguity whereas the lay user prefers crisp, simple and clear names that permit discrimination within a limited set of alternatives. Because of this, trade names and acronyms will inevitably continue to complement more rigorous terminology.

History

Polymer science and engineering probably started, as is so often the case, with Faraday (Mark, 1986), but natural polymers such as starch, proteins, and rubber- and cellulose-based substances were in use in earliest times. A series of internationally co-ordinated attempts to replace chaos by order in the nomenclature (IUPAC, 1952; IUPAC, 1966; Huggins et al., 1966; Fox et al.,1967) were reviewed by Fox (1968a), whose carefully considered comments provide an authoritative and accessible source. Further clarification of nomenclature issues will undoubtedly appear in the imminently forthcoming publication of the next volume of the second edition of the *Encyclopedia of polymer science and technology* (Wiley-Interscience).

In the last 20 years, a range of further detailed proposals for polymer naming and abbreviation have been laid down, analysed and reviewed (Fox, 1968b; Fox, 1968c; Bikales, 1969; Fox, 1970; Skolnik and Huys, 1970; Sterman, 1970; IUPAC, 1972; IUPAC-IUB, 1972; Tsuruta, 1972; IUPAC, 1973; Klempner et al., 1973; Fox, 1974; IUPAC, 1974a and b; Schultz, 1975; Sperling and Ferguson, 1975; Avrutina, 1976; Goncharuk, 1976; IUPAC, 1976; Akopov, 1977; Becker and Liebscher, 1977; Elias, 1977; Sperling, 1977; ASTM, 1978; Boyer, 1978; Russian Translations of IUPAC, 1978a and b; Kotaka, 1978; Scarito et al., 1978; IUPAC, 1979a and b; Schildknecht, 1979; Sperling and Corwin, 1979; Tsuruta, 1979; Fresenius, 1981a and b; IUPAC, 1981a and b; Carraher, 1983; Plate et al., 1983; Feuerberg, 1984; IUPAC, 1985; Pethrick, 1985; Bikales, 1986; Donaruma, 1986; Kaufman, 1986; Schwachula and Haeupke, 1986; Carraher et al., 1987; Czech translation of IUPAC, 1987; IUPAC, 1987; Kudo, 1987). The four major name types: common, source-based, characteristic group and structure-based, are discussed by Carraher, Hess and Sperling (1987), and two general trends can be discerned over this time period:-

i) an adherence to certain trade names, and/or accepted abbreviations and acronyms, many of which have been legitimized by international agreement by the International Union of Pure and Applied Chemistry (IUPAC).

ii) a movement away from source or monomer based names,

designed to reflect a particular chosen synthetic route, towards structure based names.

The largest repository of polymer information is *Chemical Abstracts* in which data are stored in two main files, a bibliographic file which contains items from the primary literature, and a registry file which identifies individual substances. These are computer stored, and the information they contain may be accessed by computer online searching, or by manual scanning of the hard copy printout.

In principle, each substance is assigned a unique registry number, the value of which is related to the time when the substance was first registered but which does not possess any significance or meaning with respect to composition or structure. Although the Chemical Abstracts Service (CAS) registry system was started in 1965, polymer information was not added until 1967 (CAS, 1983).

Despite the trend towards structure-based nomenclature mentioned above, CAS continues to use the component monomer as the primary route to polymer identification, because they feel that the structure of the final polymer may not be known, or may be insufficiently well described by the author. When they feel that a structure is soundly based, they add a supplementary structural repeat unit (SRU) based representation to the registry system, permitting user retrieval by this route in such cases. CAS are currently reconsidering their policy in this area, and are contemplating using SRUs wherever possible, and linking them to monomer-based records, thus allowing the searcher to choose his or her preferred approach (CAS, 1987). They are also considering how best to represent biological macromolecules and polymers.

A temporary expedient for systemizing polymer nomenclature, which was also used in the early days of computer information storage of organic structural information before graphics handling capabilities became common, has been the use of Wiswesser line notation (WLN) (Goncharuk et al., 1976; Pethrick, 1985).

Principles and practice of naming polymers

Monomer-based names

Elias (1977) has a useful classification of three types of definition of a polymeric substance, on the basis of which he lays down some firm rules of nomenclature. His classification is:

i) phenomenological, involving such external features as overall composition of a compound, elimination of groups during reactions, etc;

ii) molecular, based on chemical structure or on the reaction mechanism, i.e. bonds and changes occurring within those bonds;

iii) operative, including M_r degree of crystallinity, etc.

He points out that in the early days of polymer chemistry, synthetic polymers were named on the basis of the precursor monomer, which resulted in such names as polyethylene and polystyrene. Fox (1968a) takes the view that such a monomer- or source-based name is actually a contraction of 'polymer from styrene'. In other cases, the choice of name was provided by a characteristic group in the final polymer, leading to the name of polyamide for the class of polymers produced by reacting diamines with dicarboxylic acids.

A shortcoming of this simplistic method is seen in cases where more than one kind of structural repeat unit can arise from the monomer. An excellent example which he cites is the set of isomeric polymers formed from butadiene, which include the elastomeric 1,4 cis; the thermoplastic 1,4 trans, 1,2 isotactic and 1,2 syndiotactic; and the thermosetting 1,2 cyclic and 1,2 aromatic, the former being an insulator and the latter a semiconductor.

When the monomer name is multi-word, parentheses are used; an example is poly(vinyl acetate), Parentheses should also be used in such cases as poly(4-chlorostyrene); even if the substituent position were to be omitted, the parentheses would still serve to distinguish the final polymer from a polychlorinated styrene!

The disadvantages of monomer-based names can be summarized as:

i) ambiguity of structure, e.g. polybutadiene;

ii) ambiguity of synthetic route, a problem which arises because the synthetic route is not specified. Fox (1968a) considers poly(4-vinybenzaldehyde) which is actually poly(4-formylstyrene) when polymerized through the vinyl group and a polyether, truly called poly(4-vinylbenzaldehyde) when polymerized through the aldehyde group;

iii) absence of monomer character; poly(ethylene glycol) lacks the character of a glycol (Fox, 1968a), but the name would correctly describe the material $HO(CH_2CH_2)_nOH$;

iv) semantic difficulties with copolymers. Although 'polymer from hexamethylenediamine and adipoyl chloride" correctly describes poly(hexamethylene-*co*-adipoyl chloride), is monomer an accurate description of these precursors? (Fox, 1968a);

v) hypothetical nature of monomer. Vinyl alcohol does not exist, although poly(vinyl alcohol) does.

vi) chemical redundancy. Since the repeat unit in polythylene is -CH_2-, nothing is gained chemically by supposing it to be -CH_2-CH_2-, and a better name would be polymethylene.

Despite these strictures, *Chemical Abstracts* persists in preferring a monomer-based approach (CAS, 1983). They enclose the monomer structure in parentheses, using a lower-case subscript x to denote the polymer, e.g. polystyrene $(PhCH{=}CH_2)_x$, registry number 9003–53–6, although they shun the use of brackets in names. For copolymers, the dot-disconnection convention is used, whereby each monomer formula is separated by a dot, e.g. ethylene glycol-sebacic acid copolymer $[HO(CH_2)_2OH.HO_2C(CH_2)_8CO_2H]_x$, registry number 25037–32–5; no indication is made of monomer ratios.

Structure-based names

For unbranched polymers, the smallest structural repeat unit is a diradical such as methylene, -CH_2-, and a name for the polymer can sensibly be formed by putting ‘poly’ in front of the diradical name. Elias (1977) suggests that the name of the diradical is always given in parentheses, but in practice the majority of authors follow Fox (1968a), and omit the brackets for a simple alphabetical diradical name. Indeed some authors omit parentheses altogether, leading to potential confusion in such cases as polyethylene oxide! Elias (1977) further suggests that substituents are given before the name of the diradical, leading to his preference for poly(hydroxyethylene) over poly(vinyl alcohol) for -$(CH(OH)$-$CH_2)n$-.

Internationally agreed and approved usage (IUPAC, 1974a) regards structure-based definitions as primary, and process-based definitions including those based on monomer names as secondary, in contrast with the already discussed *Chemical Abstracts* usage (CAS, 1983). The IUPAC (1974a) name for the structural repeat unit (SRU) is a constitutional repeating unit.

The order in which components within the SRU are written may have significance, especially with respect to stereoregularity. IUPAC (1974a) regards $(CHRCH_2)$ as a different constitutional unit from (CH_2CHR), because they employ a convention that orients the unit, and also the polymer, from right to left. Elias (1977) deduces the orientation of the structural unit from the direction implied by the structurally based polymer name, in which substituents precede the diradical name. On this basis, poly(ethylene oxide) is called polyoxyethylene, and so the structural unit is written as (OCH_2CH_2) rather than (CH_2CH_2O). He notes that this direction is not the same as the direction of chain propagation in polyreactions.

Fox (1968a) suggests that, just as ‘poly’ implies ‘derived from’ in monomer-based names, in names based on structure the prefix 'poly' implies an open-chain repeating unit.

Hence poly(ethylene terephthalate) (PETP) has the SRU

$$\left(OCH_2CH_2O\overset{\overset{O}{\|}}{C} - C_6H_4 - \overset{\overset{O}{\|}}{C} \right)$$

even though it is derived from the cyclic acid

$$C_6H_4 \begin{array}{l} \overset{\overset{O}{\|}}{C} - O - CH_2 \\ \\ \underset{\underset{O}{\|}}{C} - O - CH_2 \end{array}$$

He admits that such an implication is not always correct; a ready example is poly(*p*-phenylene) or poly(1,4-phenylene) in which the repeat unit is $-C_6H_4-$. The PETP example provides a further example of deviation from systematic nomenclature in the retention of the early name, ethylene, rather than the systematic ethene.

A further set of uncertainties in nomenclature arises when one considers the terephthaloyl unit, -CO-C_6H_4-CO-. If one chose to adhere to a description based on the simplest possible set of sub units such as carbonyl, -CO-, and 1,4 phenylene, $-C_6H_4-$, then terephthaloyl would appear as carbonyl-1,4 phenylene-carbonyl (Elias, 1977) which is a distinctly uncommon usage. Also, if the convention whereby substituents precede the diradical name were always to be followed, then -(NH-CO-CH_2CH_2)n- would be called poly[imino(1-oxo-trimethylene)] and not polyiminocarbonylethylene (Elias, 1977). Rules have been described (Elias, 1977) to govern the sequence of diradical units within the SRU, and in summary these are:

i) highest priority comes first;
ii) priority is heterocyclic ring > linear groups containing heteroatoms > carbocyclic rings > carbon chains;
iii) substituents do not alter this order of priority;
iv) in rings, numbering of links to the main chain is so to utilize the shortest distance between the atoms involved;
v) for more than one ring of the same kind, the most substituted ring comes first; if equally substituted, the ring

with the lowest substituent position numbers comes first; and if these are equal, the higher priority is given to the ring with substituents whose initial letter comes first in the alphabet;
vi) for rings of different kinds, multi-ring systems come first, in order of number of rings; within n-cyclic units, largest rings come first; and within the same size rings, the least hydrogenated comes first;
vii) heteroatoms are taken in the order, O, S, Se, Te, N, P, As, Sb, Bi, Si, Ge, Sn, Pb, B, Hg, i.e. Periodic Table Group 6 then Group 5 and so on, descending within each group.
viii) if heteroatoms or rings are separated by aliphatic groups, the distance between the two highest priority components is minimized, so that -(-O-CH_2-NH-CHCl-CH_2-SO_2-$(CH_2)_6$-)- rather than -(O-$(CH_2)_6$-SO_2-CH_2-CHCl-NH-CH_2)- is correct because only four chain units separate O and S in the former whereas six are needed in the latter;
ix) for heterocyclic rings, the largest N-containing ring comes first, and for rings of the same size, more N causes higher priority, and if further discrimination is necessary, the ring with the largest number of other heteroatoms comes first. For rings, N comes first, the remaining elemental priority order being as in vii;
x) for double strand or ladder polymers, a colon is used to separate substituent positions, as in poly(1,4:2,3 butanetetrayl),

CH_2 — CH_2 — CH_2 — CH_2 (ladder structure)

xi) for spiro and other polymers containing tetravalent radicals, such radicals have priority over diradicals, as in poly[1,3-dioxa-2-silacyohexane-5,2-diylidene-2,2-bis(oxymethylene)],

C(CH_2—O)(CH_2—O)Si(O—CH_2)(O—CH_2) (spiro structure)

xii) end groups are usually unspecified, but if desired and known, the left-hand end group can be denoted by α and the right-hand end group by ω, as in α-chloro-ω-(trichloromethyl)polymethylene Cl-$(CH_2)_n$-CCl_3.

Trade names

Nylon is a well known generic name for polyamides, which may be divided into two classes, AB amino acids (A=amine, B=carboxylic acid) and AABB (diamine, dibasic acid). The former are described by a single number denoting the chain length of the monomer amino acid from which the polymer is derived by self-condensation, as in nylon-6 or poly(ε-caprolactam), CAS registry number 25038 54–4, caprolactam being a seven-membered ring, 3,4,5,6-hexahydro-2H-azepin-2-one, a 6-carbon atom amino acid made from cyclohexanone and hydroxylamine (Putschner, 1982).

Ring-containing intermediaries are coded by letter, e.g. I for isophthalic acid and T for terephthalic acid. For AABB materials two numbers are used, the first for the number of carbons separating the N atoms of the diamine and the second for the number of straight chain atoms in the dibasic acid, as in poly(hexamethylenediamine-*co*-sebacic acid) or nylon-6,10, sometimes written 6,10-nylon. An alternative descriptor, nylon-610 in which the separating comma is omitted, is to be deplored because of the potential confusion with the single digit AB nomenclature (Fox, 1968a).

A further list of over 300 trade names, including Cellophane, Kapton, Lucite (Perspex, Plexiglass), Mylar, Polythene, Teflon and Viton, is given by Elias (1977).

Abbreviations

IUPAC (1974a) gives a list, since extended, of almost 20 permitted abbreviations which are:- PAN, polyacrylonitrile; PCTFE, poly(chlorotrifluoroethylene); PEO, poly(ethylene oxide); PETP, poly(ethylene terephthalate); PE, polyethylene; PIB, polyisobutylene; PMMA, poly(methyl methacrylate); POM, poly(oxymethylene) or polyformaldehyde; PP, polypropylene; PS, polystyrene; PTFE, poly(tetrafluoroethylene); PVAC, poly(vinyl acetate); PVAL, poly(vinyl alcohol); PVC, poly(vinyl chloride); PVDC, poly(vinylidene dichloride); PVDF, poly(vinylidene difluoride); PVF, poly(vinyl fluoride).

Fox (1968a) rightly recommends that these should be defined at their first appearance in a paper or text. He also cautions that the same acronym may have a different meaning in a closely related field; for example, PE means polyethylene to the polymer chemist and photoelectron to the surface scientist.

Naming of stereoregular polymers

IUPAC (1974a) carefully distinguishes between the term 'constitutional repeating unit' or SRU, for example -$CH(CH_3)CH_2$-, and the configurational base units of which they are composed, in this case either $-\overset{\overset{\large H}{|}}{\underset{\underset{\large CH_3}{|}}{C}}-CH_2-$ or the non-identical $-\overset{\overset{\large CH_3}{|}}{\underset{\underset{\large H}{|}}{C}}-CH_2-$.

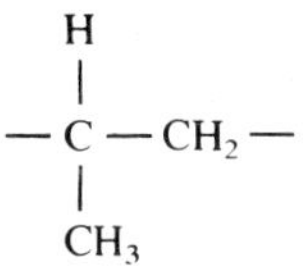

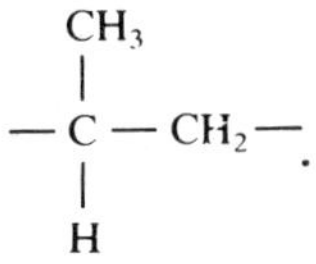

A set of adjectives to describe tacticity and related concepts has been described by Huggins et al. (1966) and Fox (1968a). For example isotactic and syndiotatic polypropylene can be distinguished both by inserting those adjectives in front of polypropylene, or by use of a hyphenated abbreviation, as in isotactic polypropylene,*it*-$[CH_2CH(CH_3)]_n$. For ditacticity, erythro and threo are distinguished by *e* or *t* preceding iso or syndio as in threo-di syndiotactic poly-2-pentene, *tst*-$[CH(CH_3)CH(C_2H_5)]_n$. Cis and trans isomerism is inserted immediately before the word stem 'tactic'.

Copolymer nomenclature

Fox (1968a) suggests that a good principle would be to generate names on a structural basis in a way that closely mimics that used for small organic molecules, but that in practice, source-based names usually have to be employed instead. The affix, *co*, is used to link the names, the component present in the largest proportion being named first, the remainder being listed in order of decreasing mass percentage. For materials such as nylons, the mass or weight proportions can also be stated. Putschner (1982) cites an example in which hexamethylenediamine, adipic acid (C_6) and sebacic acid (C_{10}) are copolymerized to give a copolymer in which the -6,6 and -6,10 components are present in the proportions 95:5, and he names this nylon -6,6–6,10(95:5).

Ceresa (1962) has assisted in the delineation of copolymer type by proposing the replacement of the conjunction '*co*' by '*alt*' for alternating, *b* for blocking and *g* for graft copolymers, yielding names such as poly(ethylene-*g*-acrylonitrile) for a copolymer formed by grafting chains from acrylonitrile onto polyethylene (Fox, 1968a). A nomenclature scheme has been proposed for graft copolymers and interpenetrating polymer networks, based on group theory concepts (Sperling and Ferguson, 1975; Scarito et al., 1978; Sperling and Corwin, 1979).

Search services and the polymer user

Information services are comprehensively covered elsewhere in this volume and the purpose of this short section is to emphasize certain constraints and strategies imposed on the polymer searcher by virtue of nomenclature-related problems. The variety of names used for the same materials can be easily seen by a casual survey of any of the 90 journals abstracted by the monthly *Polymer Contents* (Elsevier). The different styles and approaches of the polymer information provider were clearly apparent to the audience of a recent meeting on Polymer Information organized by Linford and Finch (Royal Society of Chemistry, London, June 1986, unpublished proceedings) at which presentations were given by *Chemical Abstracts*, ISI and RAPRA.

In summary, *Chemical Abstracts* provides a very broad journal coverage, but its coverage is selective (and depends on the abstractors' opinion of the degree of chemical content of the paper) and its procedures are monomer-based, which imposes the many shortcomings described above. ISI, which publishes the *Science Citation Index*, do not provide the same absolute number of journals scanned, but they do not impose artificial boundaries around the subject and, as in the case of other multi-disciplinary subjects such as tribology (Linford et al., 1974), this is an almost over-riding advantage. They can also be relied on to abstract a journal issue from cover to cover. RAPRA provide a differently constructed data base, clearly targeted to the rubber and plastics industry.

A few points in connection with *Chemical Abstracts* for online searching in addition to those already made are:

i) monomer-based nomenclature predominates and use of SRUs is largely restricted to polyamides, polyesters, polyurethanes and polycarbonates made by condensation polymerisation, and are not at present employed for polymers made by addition reactions (CAS, 1983);
ii) the order of citation from left to right in the SRU structure is determined by CAS nomenclature rules which are complicated, so that SRU-based searching of *Chemical Abstracts* is an advanced technique;
iii) end groups are usually omitted;
iv) occasionally, one polymer may appear as two distinct entries in the CAS registry file, with different registry numbers;
v) oligomers with less than 11 monomer units are separately registered, but large oligomers, of those of short but unknown chain length or with a non-integer number of repeat units, appear as polymers;
vi) telomers, i.e. low M_r polymers terminated by groups that differ from the monomer, are specially treated, e.g. the carbon tetrachloride-vinyl chloride telomer $(Cl\text{-}CH{=}CH_2)_x.CCl_4$, registry number 25655–75–8.

vii) poly(ethylene glycol) and poly(propylene glycol), which are misleadingly named as polyethylene glycol and polypropylene glycol in the searchers' manual (CAS, 1983) are registered only as SRUs;
viii) tacticity is dealt with in terms of different registry number for e.g. polypropylene, but block, graft, etc. copolymers are only so described at index guide level;
ix) post-treated polymers which have been chemically modified after polymerisation has been completed, are treated by the dot disconnection approach, with the oddity that not necessarily equal coefficients are ascribed the same symbol, e.g. the sodium salt of poly(acrylic acid), $(CH_2{=}CHCO_2H)_x.xNa$, registry number 9003–04–7;
x) polymer blends or polymer alloys are only treated in terms of the mixture components;
xi) paper chemistry of non-existent polymers on which molecular orbital or similar calculations have been carried out, can lead to the generation of SRUs, as for polynitritomethylidene, registry number 73329–66–5 for example;
xii) a dilute and unreliable sprinkling of trade or common trade names has been registered, e.g. Klason Lignin, registry number 8068–04–0, and commercially important fibres and rubbers are indexed only by such names and not under the name of the polymer of which they are composed.

Present day status and future requirements

As can be seen, the present day status is uncertain, a statement that has been continuously true for many years and continues to be valid. Because of the variety of user needs and the diversity of existing practice, it is difficult to anticipate (or indeed to wish for) a time when common names are not used in addition to systematic names; there is much to commend the concise over the elaborate. What would undeniably be useful is to proceed to a unified and systematic approach that obviates ambiguity and maximizes information content, but it seems clear that the complexities of the subject and the attraction of maintaining familiar usage will impede the attainment of this goal.

References

Akopov, A.S. (1977). 'Nomenclature of structural and polymeric analogs of phthalocyanines'. *Izv. Vyssh. Uchebn. Zaved., Khim. Khim. Tekhnol.*, **20**(9), 1281–4. CA87(26):202151g.

Albrecht, O., Laschewstry, A. and Ringsdorf, H. (1985). 'Gas permeation membranes'. *J. Membr. Sci.*, **22**, 187–197.

Anderson, C.F. and Morawitz, H. (1982). 'Polyelectrolytes'. *Kirk-Othmer Encyclopedia of Chemical Technology*. 3rd edn, (Wiley-Interscience), **18**, 495–496.

ASTM (1978). *Annual Book of ASTM Standards, Pt.35: Plastics — General Test Methods: Nomenclature* (ASTM).

Avrutina, E.A.; Andandonskaya, B.L.; Andreeva, I.N.; et al. (1976). 'Synthetic high polymers'. *Thesaurus of Polymeric Materials (Tezaurus po Polimernym Materialam)* (Gos. Kom. Standartov: Moscow, USSR), 266 pp (CA85(22):160818f).

Becker, R. and Liebscher, W. (1977). 'Contributions to the nomenclature of chemical compounds. VI. Present status of polymer nomenclature'. *Plaste Kautsch.*, **24**(12), 790–3.

Bikales, N.M. (1986). 'Polymer nomenclature'. *Encyclopedia of Materials Science and Engineering*. ed. Bever, M.B. (Pergamon), **5**, 3730–3732.

Bikales, N.M. (1969). 'Polymer nomenclature in industry'. *J. Chem. Doc.*, **9**(4), 245–7.

Boyer, R.F. (1978). 'Nomenclature for multiple transitions and relaxations in polymers'. *Midl. Macromol. Monogr.*, **4** (Mol. Basis Transitions Relaxations), 329–31.

Carraher, C.E., Jr., Hess, G. and Sperling, L.H. (1987). 'Polymer nomenclature or what's in a name?'. *J. Chem. Educ.*, **64**(1), 36–9.

Carraher, C.E., Jr. (1983). 'Polymer education'. *Polym. News*, **8**(12), 363–4.

CAS Report (1987). 'Special issue'. *CAS Report* (23), 6.

CAS (1983). *Searching polymers by structures through CAS ONLINE* (CAS, Columbus).

Ceresa, R.J. (1962) *Block and graft copolymers* (Butterworths).

Czech Translation of IUPAC (1987). 'Nomenclature of regular single-strand organic polymers'. *Chem. Listy*, **81**(3), 290–316.

Davis, G.W. and Hill, E.S. 'Polyester fibers'. *Kirk-Othmer Encyclopedia of Chemical Technology* (3rd edn, Wiley-Interscience), **18**, 531–540.

Donaruma, L.G. (1986) 'Polymer structures'. *Encyclopedia of Materials Science and Engineering*. ed. Bever, M.B. (Pergamon), **5**, 3732–3739.

Elias, H.G. (1977) *Macromolecules, 1* (Plenum).

Feuerberg, H. (1984). 'Nomenclature and symbols for polymers'. *J. Polym. Sci., Polym. Lett. Ed.*, **22**(7), 413–17.

Fox, R.B. (1974a). 'Notes on nomenclature. Naming organic polymers. II. Structure-based polymer nomenclature'. *J. Chem. Educ.*, **51**(2), 113–15.

Fox, R.B. (1974b). 'Notes on nomenclature. Naming organic polymers. I. Problems with present-day nomenclature'. *J. Chem. Educ.*, **51**(1), 41–2.

Fox, R.B. (1970). 'Structure versus reality in polymer nomenclature'. *J. Chem. Doc.*, **10**(4), 241–3.

Fox, R.B. (1968a). 'Nomenclature'. *Encyclopedia of Polymer Science and Technology*. (1st edn, Wiley-Interscience). **9**, 336–344.

Fox, R.B. (1968b). 'A structure-based nomenclature for linear polymers'. *Macromolecules*, **1**(3), 193–8.

Fox, R.B. (1968c). 'Present status on the nomenclature of polymers'. *Rev. Gen. Caout. Plast., Ed. Plast.*, **5**(3), 150–2.

Fox, R.B. (1967). 'Some problems in polymer nomenclature'. *J. Chem. Doc.*, **7**(2), 74–8.

Fresenius, P. (1981a). 'Questions about pharmacopeias. II. Remarks on polymer nomenclature'. *Dtsch. Apoth.-Ztg.*, **121**(42), 2302–7.

Fresenius, P. (1981b). 'Problems with pharmacopeias. Remarks on polymer nomenclature, especially polyethylene glycol'. *Dtsch. Apoth.-Ztg.*, **121**(9), 445–6.

Glasse, M.D. and Linford, R.G. (1987). 'Polymer structure and conductivity'. *Electrochemical Science and Technology of Polymers 1*. ed. Linford R.G. (Elsevier Applied Science) 23–44.

Goncharuk, G.P., Malakhova, E.L. and Shevyakova, L.A. (1976). 'Coding of linear polymers'. *Inf. Probl. Sovrem. Khim.*, 39-44. ed. Mikhailov, A.I. (Vses. Inst. Nauchn. Tekh. Inf., Moscow, USSR).

Huggins, M.L., Natta, G., Desreux, V. and Mark, H.F. (1966). 'Report on nomenclature dealing with steric regularity in high polymers'. *Pure Appl. Chem.*, **12**(1–4), 643–54.

IUPAC (1987). 'Use of abbreviations for names of polymeric substances. (Recommendations 1986)'. *Pure Appl. Chem.*, **59**(5), 691–3.

IUPAC (1985). 'Nomenclature for regular single-strand and quasi single-strand inorganic and coordination polymers (Recommendations 1984)'. *Pure Appl. Chem.*, **57**(1), 149–68.

IUPAC (1981a). 'Nomenclature for regular single-strand and quasi single-strand inorganic and coordination polymers'. *Pure Appl. Chem.*, **53**(11), 2283–302.

IUPAC (1981b). 'Stereochemical definitions and notations relating to polymers'. *Pure Appl. Chem.*, **53**(3), 733–52.

IUPAC (1979a). 'Nomenclature of regular single-strand organic polymers'. *Pollimo*, **3**(4), 228–44.

IUPAC (1979b). 'Stereochemical definitions and notations relating to polymers'. *Pure Appl. Chem.*, **51**(5), 1101–21.

IUPAC (1976). 'Nomenclature of regular single-strand organic polymers'. *Pure Appl. Chem.*, **48**(3), 373–85.

IUPAC (1974a). 'List of standard abbreviations (symbols) for synthetic polymers and polymer materials'. *Pure Appl. Chem.*, **40**(3), 473–6.

IUPAC (1974b). 'Basic definitions of terms relating to polymers'. *Pure Appl. Chem.*, **40**(3), 477–91.

IUPAC (1973). 'Nomenclature of regular single-strand organic polymers'. *Macromolecules*, **6**(2), 149–58.

IUPAC (1972). 'Nomenclature of regular single-strand organic polymers'. *Inf. Bull.-I.U.P.A.C., Append. Tentative Nomencl., Symb., Units, Stand.*,**29**, 31 pp.

IUPAC-IUB (1972). 'Abbreviated nomenclature of synthetic polypeptides (polymerized amino acids). Revised recommendations (1971)'. *Biochim. Biophys. Acta*, **278**(2), 211–17.

IUPAC (1966). 'IUPAC Recommendations for abbreviations of terms relating to plastics and elastomers'. *IUPAC Bull* **25**, 41.

IUPAC (1952). 'Report on nomenclature in the field of macromolecules'. *J. Polymer Sci.* **8**, 257.

Jenkins, A.D. (1984). 'The terminology for molar masses in polymer science'. *J. Colloid Interface Sci.*, **101**(1), 277.

Kaufman, S.L. (1986). 'Polymer structures'. *Encyclopedia of Materials Science and Engineering*. ed. Bever, M.B. (Pergamon), **5**, 3739–3742.

Klempner, D., Frisch, K.C., and Frisch, H.L. (1973). 'Nomenclature of interpenetrating polymer networks'. *J. Elastoplast.*, **5**(Oct.), 196–200.

Kotaka, T. (1978). 'High polymer chemistry: multicomponent polymer systems'. *Kagaku (Kyoto)*, **33**(11), 932–4.

Kudo, Y. (1987). 'Polymer retrieval system for generic polymer names consisting of both generic and specific monomer names. A prototype system for the Gazetted List of Existing Chemical Substances of Japan'. *Yamagata Daigaku Kiyo, Kogaku*, **19**(2), 147–56.

Kwolek, S. L., Morgan, P. W. and Schaefgen, J.R. (1987). 'Polymer liquid crystals'. *Encyclopedia of Polymer Science and Technology* (2nd edn, Wiley-Interscience), **9**, 1–61.

Linford, R.G., Neat, R., Glasse, M.G. and Hooper, A. (1986). 'Thermal history and polymer electrolyte structure'. *Solid State Ionics.*, **19**, 1088–1093.

Linford, R.G., Neat, R., Glasse, M.G. and Hooper, A. (1985). 'A structural model for the interpretation of composition-dependent conductivity in polymeric solid electrolytes'. *Transport-structure relations in fast ion and mixed conductors: Proceedings of the 6th Risø International Symposium on Metallurgy and Materials Science* (Risø National Laboratory, Roskilde) 341–345.

Linford, R.G., Parry, A.A. and Rich, J.I. (1974). 'Computer literature searches — a comparison of the performances of two commercial systems in an interdisciplinary subject'. *Information Scientist* Dec., 179–187.

Linford, R.G. (1973). 'Symbols and nomenclature in solid surface thermodynamics'. *J. Electroanalyt. and Interfacial Electrochem.*, **34**, 155–157.

Loening, K.S. (1981). 'Polymer nomenclature'. *Kirk-Othmer Encyclopedia of Chemical Technology*. (3rd edn, Wiley-Interscience), **16**, 28–46.

MacDiarmid, A.G. and Maxfield, M. (1987). 'Organic polymers as electroactive materials'. *Electrochemical Science and Technology of Polymers 1*, ed. Linford R.G. (Elsevier Applied Science) 67–102.

Mark, H. (1986). 'History of Polymers'. *Encyclopedia of Materials Science and Engineering*. ed. Bever, M.B. (Pergamon) **5**, 3710–3714.

Olabisi, O. (1982). 'Polyblends'. *Kirk-Othmer Encyclopedia of Chemical Technology*. (3rd edn, Wiley-Interscience), **18**, 433–466.

Owen, J. (1987). 'Ion conducting polymers'. *Electrochemical Science and Technology of Polymers 1*, ed. Linford R.G. (Elsevier Applied Science) 45–66.

Pethrick, R.A. (1985). 'Nomenclature in polymer science'. *Polymer Year Book 2*. (Harwood) 1–27.

Plate, N.A., Papisov, I.M. and Renard, T.L. (1983). 'Terminology and nomenclature of polymers'. *Zh. Vses. Khim. O-va.*, **28**(3), 301–8.

Putscher, R.E. (1982). 'Polyamides'. *Kirk-Othmer Encyclopedia of Chemical Technology*. (3rd edn, Wiley-Interscience) **18**, 328–360.

Russian Translation of IUPAC (1978a). 'Nomenclature of regular linear single-strand organic polymers. (Rules approved in 1975)'. *Vysokomol. Soedin., Ser. A*, **20**(5), 1178–97.

Russian Translation of IUPAC (1978b). 'Basic definitions of terms pertaining to polymers, 1974'. *Vysokomol. Soedin., Ser. A*, **20**(5), 1167–77.

Scarito, P.R., Corwin, E.M. and Sperling, L.H. (1978). 'Isomeric graft copolymers and interpenetrating polymer networks. A generalised nomenclature with experimental examples'. *Polym. Prepr., Am. Chem. Soc., Div. Polym. Chem.*, **19**(1), 127-32.

Schildknecht, C.E. (1979). 'Trends in naming monomers and polymers'. *Mod. Paint Coat.*, **69**(6), 41–5.

Schultz, J.L. (1975). 'Polymer nomenclature, classification, and retrieval in the Du Pont Central Report Index'. *J. Chem. Inf. Comput. Sci.*, **15**(2), 94–100.

Schwachula, G. and Haeupke, K. (1986). 'The morphology of ion exchange resins and other functional polymers. Comment on the systematisation and the terminology'. *Morphol. Polym., Proc., Europhys. Conf. Macromol. Phys.*, 17th Meeting Date 1985, 757-64. ed. Sedlacek, B. (de Gruyter, Berlin, Fed. Rep. Ger).

Skolnik, H. and Hays, J.T. (1970). 'New notation system for indexing polymers'. *J. Chem. Doc.*, **10**(4), 243–7.

Smith, P.J. (1987). 'Fluorinated ionomer membranes for use in the production of chlorine and caustic soda'. *Electrochemical Science and Technology of Polymers 1*, ed. Linford R.G. (Elsevier Applied Science), 293–334.

Sperling, L.H. and Corwin, E.M. (1979). 'A proposed generalised nomenclature scheme for multipolymer and multimonomer systems'. *Adv. Chem. Ser.*, **176** (Multiphase Polym.), 609–30.

Sperling, L.H. (1977). 'Isomeric graft copolymers and interpenetrating polymer networks. Current status of nomenclature schemes'. *Chem. Prop. Crosslinked Polym., [Proc. ACS Symp.]*, Meeting Date 1976, 217–41. ed. Labana, S.S. (Academic, New York, N.Y.)

Sperling, L.H. and Ferguson, K.B. (1975). 'Isomeric graft copolymers and interpenetrating polymer networks. Possible arrangements and nomenclature'. *Macromolecules*, **8**(6), 691–4.

Sterman, M. (1970). 'Solutions to polymer nomenclature problems in the U.S. Patent Office'. *J. Chem. Doc.*, **10**(4), 237–40.

Tsuruta, T. (1979). 'Rules for nomenclature for macromolecular compounds'. *Kobunshi Jikkengaku*, **1**, 385–415. ed. Kanbara, M. (Kyoritsu Shuppan K.K., Tokyo, Japan).

Tsuruta, T. (1972). 'Polymer nomenclature'. *Kobunshi*, **21**(243), 322–3.

Vincent, C.A. (1987). 'Polymeric electrolytes'. *Progr. Solid State Chem.* **17**, 145–261.

CHAPTER TEN

Additives and catalysts

J.A. SHELTON

Introduction

The role of the 'additive' can, in general terms, be described as a property modifier for the base polymer or resin in relation to its processability or fitness for purpose.

The main categories comprise fillers and/or reinforcements; stabilizers (both heat and light); plasticizers; fire retardants; colourants; antistatic agents; lubricants and blowing agents. Catalysts must be considered as a category apart.

High polymers (plastics) are highly sensitive to heat and/or light as a consequence of their chemical structure and carbon 'backbone'. This inherent sensitivity has specific advantages and allows plastics to be formed into shapes etc., which are often outside the scope of other materials, by the use of relatively low temperatures and pressure in, for example, the injection moulding or extrusion process. It also has distinct disadvantages in that it imposes limitations on their end-uses. Without additives, plastics would be wholly unsuitable for a large number of current applications. Additives have upgraded plastics to the extent that they have become 'engineering or structural materials' in their own right compared with the early plastics which were regarded essentially as decorative materials, at least insofar as the thermoplastics were concerned.

The requirements, in terms of information relating to the additives sector, can generally be subdivided into 'technical' or 'commercial'. The catalogue of available compounds and materials is being added to almost daily, and although the performance characteristics may be only marginally improved, there can be cost-performance advantages associated with the newer products which is an all-important feature of any additive compound. There is also a need to know what is available and from whom, and the first publication which provides a comprehensive guide to commercial products has recently been published by Noyes Publications in the U.S. under the title *Plastics Additives — an industrial guide*, Flick (1986).

This describes almost 3,400 additives for plastics which are currently available to industry. It has been compiled from information received from numerous industrial companies and other organisations on their products, and only the most recent data has been included. Each section covers a specific additive class, sub-divided into individual products/suppliers and their products. The number of 'raw materials' covered in each section varies according to the overall commercial importance of the end-use; for example, there are some 605 entries under FILLERS AND EXTENDERS but only 9 under ANTIFOGGING AGENTS. It does not cover CATALYSTS. Readership will include technical and managerial personnel involved in all facets of plastics utilisation. It is limited to materials of American origin, but nevertheless forms a concise and useful reference work on this subject, although it will require updating every two years or so to sustain its value. Strictly speaking, the title should be 'Additives for Polymers', since 'Plastics Additives' implies that additives are made of plastics.

There is only one abstract publication devoted entirely to additives which provides a current awareness up-date on a monthly basis, and that is *Additives for polymers*, (Elsevier — Technology Publications), Oxford). This is compiled and edited by Fulmer Yarsley (Redhill), an independent company specializing in contract research and development in the polymer and non-metallics field, who founded this publication some 15 years ago. It gives concise information under the following sections: a) technical notes, b) materials, c) applications, d) U.K. and U.S. patents, e) markets, f) company news, g) books/publications and h) forthcoming calendar events. It covers material and commercial developments world-wide, and the markets section provides a source reference to the in-depth market studies and reports on additives which are now published with increasing regularity.

Fillers and reinforcements

A good example of the important role of additives is polypropylene which, in its unmodified form, is used mainly in non-demanding applications, but combined with a filler, such as talc or chalk, it is transformed into a compound suitable for the construction of critical end-uses in a number of industrial sectors, particularly the automotive sector.

This applies to the majority of the 'commodity' plastics, and selection of the filler and/or reinforcement, and the polymer/filler ratio has become a science of considerable importance. Fillers were at one time regarded as a method of cheapening the product, but in fact the energy requirements to compound a filler and polymer are both high and costly, and this cost is handed on in the form of a premium for filled compounds.

Selection of the filler in relation to end-uses is critical, and polypropylene may once again be used by way of example when considering the selection of mineral fillers. Although talc and chalk are, to the layman, very similar in format, they behave differently when used in plastics, and properties such as impact and tensile strength will depend upon the selection criteria. The value of the filler/reinforcement has really gone full circle from a possible way of reducing the amount of base polymer, to the highly effective use of the polymer as a matrix material in order to utilize the filler/reinforcement to its fullest potential.

The susceptibility of plastics to temperature is markedly reduced by the scientific use of fillers since they allow the inherent so-called 'creep' characteristics of the unmodified polymer to be considerably lessened.

It is the mineral fillers which are of major interest to the plastics industry in terms of particulate or powdered forms. In the fibrous form, glass and carbon — often used in synergistic combination — are prominent. Glass is used in microsphere form (both hollow and solid) and the regular geometry of these spheres enables the precise tailoring of compounds.

Mention must be made of the specialist, and therefore high-cost, fibres now commercially available. These include fibres of boron, silicon carbide, aluminium oxide, and the established aramids. These are of particular value in meeting the exacting demands of the aerospace and transport industries, and also the sports and leisure sectors.

One recent publication, Bhattacharya (1986), provides an excellent overview of metallic and non-metallic fillers in plastics, linking the conferred properties with specific end-uses. Its primary object is to provide a current state-of-the-art in respect of metal-filled polymer systems or composites in the style of a monograph.

A further monograph, Wake (1971), deals with the properties of the various mineral and other powders and some fibres and fibrous materials commonly in use in plastics at that time. It refers to both fillers and reinforcements under the colloquial term 'fillers'. It is a well indexed publication and was considered to be one of the major sources of reference in the 1970s. It deals with the physical and chemical structure, preparation, and sources of fillers used in rubbers and plastics, and contains valuable data in tabulated form.

In Titow and Lanham (1975), the term 'reinforcing fillers' is preferred when discussing fibre-reinforced thermoplastics. It covers the technology of 'reinforced' thermoplastics and some related commercial aspects, with special reference to the nature and properties of the industrially important members of this class of materials as well as their main constituents and methods for their production. The index is adequate, but could have been somewhat more comprehensive particularly as this work is a valuable reference source on the properties of fibrous fillers and on test methods associated with their evaluation. It also contains patent literature, both U.K. and U.S., relating to reinforced thermoplastics per se.

The index in Ritchie (1972) which covers fillers, stabilizers and plasticizers, is good, and is to the standard generally associated with publications sponsored by The Plastics Institute in the U.K. (now the Plastics and Rubber Institute). The section on fillers is sub-divided into two main chapters, one on their structure and one on their applications. Both chapters are the work of King (Ciba-Geigy).

The field of advanced plastics composites and related fibre technology is covered in a very comprehensive manner by Lubin (1969). Published under the SPE Polymer Technology Series, this handbook is a guide to engineers, designers, and technicians on this complex subject. It was the first publication to list all the available glass formulations, their properties and finishes. It also lists all the major processing techniques. It contains the work of 32 experts within this industry and is sub-divided into raw materials, which includes chapters on the speciality and high-temperature resistant thermoset resins as well as glass-filled thermoplastics, and the various fibre forms used, from glass through to boron and graphite whiskers. The second section details processing methods for such composites. The third covers applications.

An international conference devoted specifically to 'Fillers' was held in the U.K. in 1986, sponsored jointly by the Plastics and Rubber Institute and the British Plastics Federation. A paper given by Trubshaw and Christie *Fillers in perspective* highlighted the likely developments within the immediate future, including a) optimized particle size distribution and hybrid systems, b) improved surface-coupling techniques, c) tailoring of the matrix polymers and d) improved compounding conditions and equipment. Although these developments are generally associated with the thermoplastics, it should be emphasized that such research will also be pertinent to elastomers and the thermosets.

Fillers and their effect on properties and their cost-effectiveness in PVC are discussed by Titow (1984). This stresses the importance of calcium carbonate in the vinyl sector.

Stabilizers

There are essentially three main categories; 1) heat stabilizers (which are generally associated with the processing of PVC), 2) light stabilizers (which may alternatively be described as ultraviolet stabilizers or absorbers or under the general term of antidegradants), and 3) antioxidants (which protect against oxidation).

All three categories play a decisive part in plastics technology. Their primary function is to protect polymers from a) discolouration or decomposition during the processing stage, and b) from environmental degradation in the moulded or finished form, particularly in outdoor applications, e.g. PVC window frames. At the present time, there are about 20–25 major companies supplying stabilizers and stabilizer systems within W. Europe with a small number of speciality producers.

Heat stabilizers are, as stated, used with PVC and other chlorine-containing polymers, and the main types include barium-cadmium soap blends, calcium-zinc, lead salts and soaps, organotins, organo-antimony derivatives, organo-phosphites, and epoxidized oils and esters. Of these, the organotins are currently showing the highest growth rates since they are extremely efficient and colourless. Barium-cadmium soap blends are the most widely used of the established stabilizers and are sometimes used in conjunction with zinc. Lead salts and soaps are the oldest of all, but are the most toxic and therefore raise environmental health problems. The so-called primary antioxidants such as hindered phenols and bisphenols and amines protect the polymer against oxidation since they inhibit free-radical chain propagation, and are usually used in a synergistic sense with secondary antioxidants such as organo-phosphites and thioesters. The phosphites offer protection during processing and the thioesters against environmental ageing.

The current trend is towards higher molecular weight compounds which have a greater permanency. As processing temperatures increase, so the demands placed upon antioxidant formulations increase, and in common with other 'additive' compounds, the practice of 'tailoring' to optimize cost/performance characteristics, or synergistic blending, is evident.

Light or ultraviolet (UV) stabilizers include metal salts of nickel, aryl and acrylic esters etc., but 2-hydroxybenzophenones are the largest and most versatile class of UV stabilizers. They are compatible with many other UV absorbers and are particularly effective in the high-volume polyolefines sector. Hindered amine light stabilizers (HALS) are now significant in terms of usage. They can be used synergistically with phosphites. The increasing demand in the so-called 'engineering plastics' field and especially the high-temperature resistant materials, has placed a greater emphasis on efficient stabilizer systems with long-term effectiveness.

The energy of UV light in the 3000–4000 Å region is capable of splitting most chemical bonds, and although many components will absorb UV light, a large number are unsuitable for use in plastics, especially where colour is important; for example, carbon black is an excellent UV absorber but its colour limits its widespread use. White pigments such as titanium dioxide and zinc oxide have been considered as UV absorbers, but their effectiveness is thought to be due to light scattering or dispersion in thermal terms rather than to absorption.

Of all the polyolefines, polypropylene is the most sensitive to UV light which rapidly leads to severe reduction in gloss or to surface 'chalking', and a general deterioration in mechanical properties. Of the nearly 3,100 tonnes of light/UV stabilizers used in the EEC in 1986, about 40% was used in polypropylene.

The monograph edited by Ritchie (1972) describes the mechanisms of stabilization, and covers antioxidants, heat stabilizers, and UV absorbers. It then goes on to discuss the stabilization of individual polymer types and classes, with the emphasis on PVC and the polyolefines. It also mentions a further important aspect of stabilization, that of the susceptibility of plastics to bacterial and fungal attack. This subject has received very little in the way of 'technical publicity' and Ritchie devotes only one page in outlining the ways in which such attack occurs, and the anti-microbial agents that are effective additives. These include chlorinated phenol compounds, copper 8-hydroxyquinolinolate, an organic-mercury compound, *N*-trichloromethylthio-4-cyclohexane-1,2-dicarboximide, and bis(tributyl tin) oxide and its salts.

Biodegration (along with photodegradation) was considered an effective way of disposing of waste plastics packaging many years ago, and Professor Scott developed compounds in which the bio- or photo-degradation process could be more or less controlled. Although several patents were granted to Scott, the viability and cost of these compounds were questionable in the commercial sense.

There is a noticeable lack of textbooks devoted to the subject of polymer stabilization, and even less on specific stabilizer systems. The monograph edited by Neiman (1965) on the ageing and stabilization of polymers retains its value in view of its in-depth academic study of the subject. It is a translation of studies carried out in the Soviet Union, designed to create a scientific basis for the critical selection of effective stabilizers. The translations were made by a large group of authors, each one a specialist in polymer stabilization. Chapters deal with the basic mechanisms of thermo-oxidative degradation and stabilization, the role of radicals as inhibitors or oxidative processes, the synthesis of stabilizers, and the mechano-chemical processes in highly elastic polymers. Other chapters are devoted to individual classes of polymers. It contains an extensive bibliography of periodicals and patents.

Titow (1984) provides a background view on the inherent instability of PVC and the need for stabilization, with reference to both thermal and photochemical degradation. Stabilizer evaluation is considered in some detail in one chapter, and a further chapter considers the range of stabilizers available from five UK suppliers with details of typical applications. Stabilization and ageing of polymers have, however, been the subject of numerous papers and articles published in the technical press and a number of the more recent are listed at the end of this chapter.

Fire retardants

Fire retardancy is a complex subject, but essentially it concerns the reduction of the inherent flammability of certain plastics. The fire retardant — in the form of an additive compound — may also act as a smoke suppressant, which is as important as the problem of flammability in those polymers which generate harmful or toxic products during combustion. This is particularly the case with the polyurethanes, and the essential need to retard this class of materials (especially in the foamed form), has been accelerated by various government standards and regulations.

Since all additives confer a cost premium on the product, the aim is to optimize the use of a fire or flame retardant chemical to achieve, at reasonable cost, the greatest efficiency without detrimental effect on the mechanical or physical properties of the base polymer. In terms of chemical composition, fire retardants are usually sub-divided into two groups — organic and inorganic. Current emphasis is on the development of non-halogenated alternatives, and also upon compatible halogenated alternatives, dispersions that reduce dusting, synergistic combinations, and fire retardants designed specifically for use in polyurethane foams.

In the inorganic group, aluminium hydrate is the largest volume fire retardant used in W. Europe. It is low in cost and, unlike the halogenated compounds, its fire retardancy is based upon the property of splitting off chemically bound water at temperatures above 200°C. The major proportion of inorganic retardants are more acceptable than the organics from the environmental point of view, but they are less effective and must be used in larger quantities which may adversely affect the mechanical properties etc. of the end-product.

The inorganics include inorganic phosphorus products, aluminium hydrate, magnesium hydroxide, boron/zinc compounds, and antimony oxide; the latter two products being used primarily as synergists.

The organics include organobromine compounds, organochlorine compounds, organofluorine compounds, organophosphorus compounds, brominated organophosphorus compounds, and chlorinated organophosphorus compounds.

In general the use of a synergist, and particularly of antimony oxide, allows the user to reduce significantly the quantities of the more expensive compounds used in a given formulation.

As previously mentioned, various government standards and regulations, including those of the EEC, are the major contributory factor insofar as supply and demand of fire retardants are concerned; without these regulatory requirements the use of fire retardants in plastics would be minimal. Growth rates for fire retardants within W. Europe in the foreseeable future have been estimated at between 2–4% per year, with the organobromine compounds showing the highest growth potential at the present time. There appears to be no direct equivalent replacement product for this compound on the horizon. The plastics user-industries are by far the largest consumers of fire retardant chemicals. In general, most W. European suppliers concentrate more or less on one major chemically-based product or product group.

There appears to have been very little published by way of textbooks devoted specifically to this subject, with the exception of a comprehensive work by Troitzsch (1984) entitled *International plastics flammability handbook*. It is a comprehensive document of about 500 pages. It covers the fundamental aspects of combustion of a wide range of plastics materials, and includes an extensive bibliography. Topics include thermal degradation, ignition, flame spread and generation of smoke and other decomposition products. The principles of flame retardants and smoke suppressants are reviewed with reference to their range of composition and application to the more important plastics groups. The 'methodology' of fire testing is analysed and the wide range of current approaches to the problem are examined against a sound basic hazard assessment technique. The major part of the volume is devoted to a review of fire test methods adopted in nearly 20 countries and the regulations to which they apply, covering buildings, transport, engineering, textiles and building contents. The sections are systematically defined and subdivided and test methods and classifications tabulated, often more clearly and concisely than in their original form. A chapter on methods of measurement of smoke and products of decomposition in fires, together with a review of their toxicological effects, is included in the final section concerned with secondary fire effects. Finally, useful appendices list terms and definitions, standards, organisations, etc. and commonly used abbreviations.

The reader will find that fire retardancy is well covered in books and publications devoted to the general state-of-the-art of additives for plastics. In the *Plastics additives handbook* edited by Gachter and Müller (1983) a chapter is devoted to fire retardant agents. Important fire retardants are divided into halogenated compounds, synergists and smoke suppressants. Detailed examples of formulations for many fire retardant plastics are given, and product trade names and manufacturers are also listed.

Additives for plastics edited by Seymour (1978) devotes a section to fire retardancy which discusses a) the general mechanism by which fire retardant additives function, b) selected polymer systems containing them, and c) the current concerns about their potential toxicity and environmental impact.

Some of the most valuable and up-to-date information is to be found in recent conference papers which review both the technical and commercial progression of fire retardants and retardancy. A conference devoted entirely to 'Flame retardant chemicals — markets and trends' was held in London in 1986. A paper by Schuller-Gotzburg presented at this conference gave an extremely comprehensive overview of flame retardants within the European sector. Apart from concise technical data on their chemistry, it contained considerable data in tabulated form on the market situation and likely prospects. The conference also included papers on specific types of retardants, such as thermally-active and vapour-active ones, boron compounds, and chlorinated organics.

Plasticizers

A plasticizer is generally defined as a substantially non-volatile, high boiling substance which, when added to another material, changes certain physical and chemical properties of that material. The plasticizer and plasticized material are held together by intramolecular forces (secondary valences). The term 'plasticizer' can, however, mean different things to different users, and definitions are at some variance with one another, but the above definition is essentially accurate in the field of plastics. Certain plasticizer types will also be referred to in the literature as 'extenders' since they can be regarded as diluents. Plasticizers as we know them today in the polymer sector may either be monomeric or polymeric in nature. Recent developments have been directed towards the rationalisation and/or optimisation of raw materials and processes associated with the established types. The possibility of new plasticizer compounds becoming commercially viable is remote.

The phthalates represent some 75% by weight of all plasticizer usage, and have been used extensively in PVC since the 1950s, but present day markets, especially in the UK, show something of a decline due to the need for more specialisation in terms of, for example, high and low temperature resistance, etc.

The phosphate plasticizers are in a much more stable position, although their cost is two and a half times that of the phthalates. There are only three producers of phosphates in W. Europe. These plasticizers also impart fire retardant properties, so they find use in PVC belting used in coal mines, etc.

The trimellitate group of plasticizers have shown rapid growth since they became commercial some 10 years ago. They are essentially esters based on trimellitate anhydride and are used as primary plasticizers in PVC compounds intended for use at high temperatures or where resistance to aqueous extraction is essential. Another important use is in PVC sheeting used in automobile interiors where they impart 'anti-fogging' properties. They also cost around two and a half times the price of conventional phthalates.

The polymeric plasticizers are mainly of the polyester type, and cost between two and three times as much as the phthalates. Their prime use is in PVC especially in food packaging applications, since they improve the resistance to migration and extraction.

Sears and Darby (1984) describe in detail the role of plasticizers and plastication in polymers, particularly PVC. The effect of plasticizers on physical properties is considered in the light of plasticizer type and content. Theories and methods of predicting the compatibility of plasticizers with various resins and plasticizer action theories (lubricity, gel, free volume) are the main unifying themes. A unique feature is the coverage of permanence or endurance of plasticized PVC in use. Much previously unpublished data is included.

Despite its age, the publication by Mellan (1961), provides the reader with a sound thesis on the behaviour of a plasticizer, which is dependent upon its inherent physical properties and chemical structure, and how such properties as vapour pressure, viscosity and melting point affect the molecular aggregate which is formed with the polymer. Other subjects of importance covered by Mellan are efficiency and retentivity, flexibility, tensile strength, modulus of elasticity, transition temperatures, internal plasticization, shrinkage, creep behaviour, low temperature performance, flame resistance, evaluation and second-order transition, volatility, diffusion migration, resistance to extraction, stiffening, heat and light stability, viscosity, toxicity, etc.

The *Plastics additives handbook*, edited by Gachter and Müller (1983) devotes chapter 5 to plasticizers, and discusses the history, definition, quality requirements and markets. Incorporation techniques and effects on gelation, hardness, mechanical properties and electrical properties are examined. The current range of plasticizers are described under their respective chemical groups. The handbook lists trade names and manufacturers, and also gives the abbreviations as prescribed in DIN 7723.

Titow (1984) discusses the various aspects of PVC plasticization and the current understanding of plasticization in some detail in the revised version of *PVC technology*. A reference chapter describes the various plasticizer types giving trade names and suppliers, whilst a further chapter discusses end-use properties imparted by the respective types.

Colourants

There are essentially two methods for the colouring of plastics: surface colouration (painting, printing, dyeing) and mass colouration. The major proportion of plastics are coloured by the latter method. Colourants, as used in plastics, can be subdivided into organic and inorganic types. The organics, both dyes and pigments, are based upon carbon chemistry whereas the inorganic pigments are mainly metal compounds. Compounding still remains the most accurate colouring technique and is particularly valid for small lots. One of the advantages of compounding is that the density of colour is higher than with other techniques, and in some instances it will produce a better dispersion of the colour than is possible using the so-called masterbatch technique. The major share of the colouring business today is achieved by using masterbatches. These are either 'polymer specific' or based upon what is termed a universal carrier system. The bulk of the masterbatch business is in black and white formulations and a high proportion of this goes into the film market. One estimate suggests that the masterbatch technique accounts for around 60% of the total colouring business in the UK.

The lead-based and chrome-based pigments are hardly used today except in applications where they are unlikely to come into regular contact with the public. Despite the frequently aired discussions on cadmium-based pigments, colouring companies see no replacement in sight for these products, particularly the reds. These cadmium pigments are bright, strong and very stable and when incorporated into the plastics material, they present no danger to the consumer. They do, however, need careful handling.

There are always hazards and problems associated with powdered pigments, and at one time it was thought that liquid colourants would make tremendous in-roads into the colouring business, but in reality they have struggled to gain market acceptance, since dry colouring is usually considered the most cost-effective method.

Factors that affect the choice of a colourant include: 1) heat fastness, sublimation fastness, 2) light and weathering fastness, 3) non-migratory, non-bloom or plate-out, 4) dispersion, 5) toxicity factors, 6) special requirements, such as fluorescence, alkali fastness, etc., and 7) cost/performance.

Among recent developments in the colour field is the introduction of hyperdispersants. Originally for non-aqueous systems in the printing ink field, their use has extended into the plastics field and they are now well established and expected to progress even further in future years. They result in much higher colour/solids concentrations.

The two most important factors influencing the development of pigments and dyes for plastics are 1) environmental aspects, with attention being given to those materials considered to be a health risk, and 2) exterior durability, e.g. improved light and weather fastness, and resistance to stress cracking, etc. The newer polymers (including the engineering plastics) are placing a demand on pigment and dye manufacturers in view of the high processing and end use temperatures associated with some of these plastics. The market for polycyclic polymer dyes has grown dramatically in the past five years, and this growth is likely to continue. Organic yellows, oranges and reds will continue to make in-roads into cadmium, lead and chrome markets.

Ahmed (1979) covers both theory and practice in *Colouring of plastics* in an easy-to-read format. It deals with the various aspects of colouring in a logical manner, from a description of colour and appearance through to the selection of colourants and their dispersion and compounding into plastics. It also contains a useful chapter on regulatory requirements in the USA with regard to Food and Drug Administration approval for 'food contact' uses etc. The surface colouration of plastics is given brief coverage. The book is heavily based upon material presented at ANTEC and RECTEC conferences sponsored by the Society of Plastics Engineers in the USA, which is a reliable source for any textbook.

Other additives

These include antistatic agents; antimicrobial agents; biocides; processing aids and lubricants; blowing agents; coupling agents; crosslinking agents or initiators; dispersion aids; impact modifiers; and release agents.

Antistatic and antimicrobial agents and biocides are, for obvious reasons, used only when such properties are required. The antistatics may be of the internal (incorporated) type, or alternatively applied externally. One of the largest markets for antimicrobial agents has been in swimming pool liners and roofing membranes made in vinyl materials to prevent the formation of mildew and algae.

Processing aids and lubricants are self-explanatory, the processing aids being particularly associated with the extrusion process.

Blowing agents may be of the 'chemical' or 'physical' type and function by producing discrete gaseous 'particles' within the polymer melt to produce low density products which may be in the form of mouldings, film or sheet.

The 'chemical' blowing agents (CBAs) are dominated by the azo types but non-azos based upon sodium borohydride or sodium bicarbonate/citric acids are under active development. The major physical blowing agents (PBAs) are methylene chloride and chlorofluorocarbons, but both of these products are under attack on health hazard (methylene chloride) and environmental pollution (chlorofluorocarbons) grounds. These blowing agents produce low density products, alternatively referred to as 'Foams', 'Expanded Plastics' or 'Cellular Plastics'. Their importance has increased with the advent of 'Structural Foams' based upon thermoplastics, and 'Reaction Injection Moulding' based almost entirely on the urethanes at the present time.

Curing agents or crosslinking agents effectively cross-link the molecular chains to produce an infusible three-dimensional structure which effectively becomes a thermoset structure.

Catalysts

Additives which act as catalysts are used mainly in the production of urethanes, where the following chemical types are in current use: 1) amines, 2) isocyanurates, and 3) tin.

Traditionally based upon the use of toluene di-isocyanates (TDI) or diphenylmethane di-isocyanates (MDI) as TDI/MDI blends, flexible foam applications have begun to focus on all-MDI materials systems which can result in higher productivity and improved foam properties. Amine catalysts are now being tailored specifically for use with the all-MDI systems, and some amines can be used in a synergistic manner with other amine-blowing catalysts. A new generation of polyols that supplement or totally replace traditional tins or amine catalysts, are autocatalytic, effectively replacing part of the polyol component, and can produce processing and property advantages in rigid boardstock formulations.

However, the established catalysts for flexible polyether- and polyester-based foams retain their performance in view of their cost-effectiveness, despite one or two associated problems such as odorous volatiles. The amines include N-ethylmorpholine, dimethylbenzamine, triethylamine, etc., but blends of amine catalysts can also be used effectively as alternatives to these compounds. One of the primary aims in catalyst development (apart from that of cost-effectiveness) is to achieve total reactivity during moulding so that no residuals remain to cause odour or toxicity problems.

Some of the earlier catalyst types are now being further exploited as restrictive patents begin to expire.

In view of the current activity in the urethane catalyst sector, the reader is well advised to consult the polymer-related journals, such as the monthly *Modern Plastics International*, (McGraw-Hill), the monthly *Journal of Cellular Plastics* (Technomic Publishing), and the bimonthly *Cellular Polymers* (Elsevier), which periodically provide a review of present developments in this field. The monthly *Modern Plastics International* (McGraw-Hill) includes catalysts for urethanes in its annual review of the additives sector in the USA normally published within the Sept.–Oct. issues of each volume. In common with other additives sectors, there is considerable emphasis on the tailoring of catalyst systems to specific production or product needs. The blending of tin catalysts, for example, can increase production rates in the moderately sized carousel lines by a half, compared with amine catalysts as used on the large so-called 'race-track' production lines.

In the monthly *Modern Plastics International* (Nov. 1986), (McGraw-Hill) the importance of the hindered amines is discussed in relation to the automotive sector, which is constantly demanding improved performance formulations effectively to utilize robotic production facilities which require the fine-tuning of amine-based reaction injection moulding systems to leave less flash in the mould and post-cure more rapidly in jig fixtures. Hindered amine catalysts (HALS) are proving to be more capable than conventional catalysts in tailoring reactivity timing, by incrementally slowing reactions to meet the specific end properties required in a moulded part.

Both the monthly *Journal of Cellular Plastics* (Technomic Publishing) and the bimonthly *Cellular Polymers* (Elsevier) carry papers emanating from academic and industrial sources. Further valuable sources of information are the Annual Technical/Marketing Conferences sponsored by The Society of the Plastics Industry (SPI) and The Society of Plastics Engineers in the U.S.

A paper presented by Taylor, G.A., Bye, M.L., Takahashi, A., and Nicholais, C.A. at the 29th Annual Technical/Marketing Conference of the SPI (Oct. 1985) described catalytic amine crosslinkers for polyurethanes, and discussed the performance of such crosslinking agents in the so-called high resilience (HR) foams illustrating their potential advantages. The materials detailed in this paper were the sterically hindered diamines.

The reaction injection moulding processes (RIM and RRIM) use a wide spread spectrum of catalyst systems including tin/amine combinations, co-catalyst systems such as tin/tin, amine/amine, as well as single amine and single tin catalysts. Polyols with built-in amine functionality are also used in conjunction with a tin/tin catalyst or tin with a very low ratio of an amine.

Catalyst optimisation is particularly evident in the flexible slabstock sector, in methylene chloride-blown materials, and in water-blown high resilience foams, etc. Catalysts now exist which allow higher levels of methylene chloride blowing agent to be used in flexible stock and which also overcome the problem of low blowing agent levels associated with the unmodified tertiary amine catalysts.

References

1. Textbooks

Allen, N. S. (1983). *Degradation and stabilization of polyolefines* (Elsevier).

Ahmed, M. (1979). *Colouring of plastics — theory and practice* (Van Nostrand Reinhold).

Aseeva and Zaikov (1986). *Combustion of polymer materials* (Hanser).

Bhattacharya, S. K. (1986). *Metal-filled polymers — properties and applications* (Marcel Dekker).

Burns, R. (1983). *Polyester moulding compounds* (Marcel Dekker).

Calbo, L. (1987). *Handbook of coating additives* (Marcel Dekker)

Cheremisinoff, N. P. and Cheremisinoff, P. N. (1978). *Fiberglass — reinforced plastics deskbook* (Ann Arbor Science Publishers).
Davis, A. and Sims, D. (1983). *Weathering of polymers* (Elsevier)
De Dani, A. (1960). *Glass fibre reinforced plastics* (Newnes).
Donnet, J. B. and Bansal, R. C. (1984). *Carbon fibres* (Marcel Dekker)
Evans and Karpel (1984). *Organotin compounds in modern technology* (Elsevier).
Flick, E. W. (1986). *Plastics additives — an industrial guide* (Noyes Publications).
Folkes, M. J. (1983). *Short fibre reinforced thermoplastics* (Wiley)
Frisch, K. C. and Klempurer, D. (1984). *Advances in urethane science and technology* (Technomic).
Gachter, R. and Müller, H. (1985). *Plastics additives handbook* (Hanser).
Garner, D. P and Stahl, G. A. (1983). *Effects of hostile environments on coatings and plastics* (American Chemical Society).
Gould, R. F. (1968). *Stabilisation of polymers and stabilizer processes* (American Chemical Society).
Grassie, N. (1985). *Development in polymer degradation.* Vol. 1–6 (Elsevier).
Grassie, N. and Scott, G. (1985). *Polymer degradation and stabilisation* (Cambridge University Press).
Harris, B. (1983). *Developments in GRP technology*, Vol. 1. (Elsevier).
Hawkins, W. L. (1984). *Polymer degradation and stabilisation: properties and applications,* Vol. 8 (Springer-Verlag).
Hilyard, N. (1982). *Mechanics of cellular plastics* (Applied Science).
Holister, G. S. and Thomas, C. (1966). *Fibre reinforced materials* (Elsevier).
Jellinek, H. (1983). *Degradation and stabilisation of polymers*, Vol. 1. (Elsevier).
Katz, H. S. and Milewski, J. V. (1978). *Handbook of fillers and reinforcements for plastics* (Van Nostrand Reinhold).
Kuryla, W. C. and Papa, A. J. (1978). *Flame retardancy of polymeric materials*, Vol. 4. (Marcel Dekker).
Lubin, G. (1969). *Handbook of fibreglass and advanced plastics composites* (Van Nostrand Reinhold).
Mellan, I. (1961). *The behaviour of plasticizers* (Pergamon).
Nadeau, H. G. (1980). *Fire property data — cellular plastics* (Technomic Publishing).
Neiman, M. B. (1965). *Ageing and stabilisation of polymers* (Consultants Bureau).
Owen, E. D. (1984). *Degradation and stabilisation of PVC* (Elsevier).
Pritchard, G. (1984). *Developments in reinforced plastics*, Vol. 3. (Elsevier).
Pritchard, G. (1984). *Developments in reinforced plastics*, Vol. 4. (Elsevier).
Ritchie, P. D. (1972). *Plasticizers, stabilizers, and fillers* (Iliffe).
Schnabel, W. (1981). *Polymer degradation.* (Hanser).
Sears, J. K. and Karby, J. R. (1984). *The technology of plasticizers* (Society of Plastics Engineers).
Seymour, R. B. (1978). *Additives for plastics*, Vol. 1 and 2. (Academic Press).
Sheldon, R. P. (1982). *Composite polymeric materials* (Elsevier).
Shlyapintokh, V. (1984). *Photochemical conversion and stabilisation of polymers* (Hanser).
Stepek, J. and Daoust, H. (1983). *Polymer properties and application — additives for plastics* (Springer-Verlag).
Struik, L. C. E. (1978). *Physical aging in amorphous polymers and other materials* (Elsevier)
Titow, W. V. and Lanham, B. J. (1975). *Reinforced thermoplastics* (Applied Science).
Titow, W. V. (1984). *PVC Technology.* 4th Edn (Elsevier).
Troitzsch, J. (1983). *International plastics flammability handbook* (Hanser).
Wake, W. C. (1971). *Fillers for plastics* (Iliffe).

Wendle, B. C. (1985). *Structural foam — a purchasing and design guide* (Marcel Dekker).
Wypych, J. (1986). *Polyvinyl chloride degradation* (Elsevier).
Yescombe, E. R. (1976). *Plastics and rubber — world sources of information* (Applied Science).

2. Publications

Additives for Polymers (Monthly) (Elsevier/Fulmer Yarsley).
Carbon and High Performance Fibres Directory, Vol. 3. (1985) (Pammac Directories).
Encyclopedia of Composite Materials and Components (Technomic Publishing).
Encyclopedia of Polymer Science and Engineering, 2nd Edn (1985) (Wiley).
European Plastics Buyers Guide (Annually) (IPC Industrial Press).
Fire Retardant Products and Their Uses (1983) (Chemical Industries Association).
Modern Plastics Encyclopedia (Annually) (McGraw-Hill)
Plastics Edition 8 (1986) (DATA).
Plastics Industry Directory (1984) (MacLaren Publishers/British Plastics Federation).
Plastics World Directory (Annually) (Cahners).

3. Study reports

Battelle (1984). *Polymer blends.*
Battelle (1986). *Carbon fibre reinforced polyphenylene sulphide.*
Business Communications (1984). *Plastics fillers and extenders.*
Business Communications (1984). *Plastics UV stability.*
Chem Systems (1985). *Additives for Styrenic and Engineering Plastics.*
Chem Systems (1986). *Antidegradants in plastics.*
Eldib Engineering and Research (1986). *High strength polythene fibres.*
Freedonia Group (1986). *Demand for plastics additives.*
Frost and Sullivan (1986). *Antioxidants in W. Europe.*
Frost and Sullivan (1986). *European additives market.*
Frost and Sullivan (1987). *Heat and light stabilizers in the EEC.*
Industrial Aids (1986). *Thermoplastic composites.*
Information Research (1985). *Demand for additives by the European paint industry over the next decade.*
Kline (1985). *Advanced polymer composites.*
Kline (1985). *Filler pigments in W. Europe.*
Kossoff and Associates (1984). *Compounding.*
Ontario Research Foundation (1985). *Fine particles and fillers — overlooked opportunities.*
Oxenham Technology Associates (1984). *Catalysts for polyolefines.*
Phillip Townsend Associates (1987). *Thermoplastics compounding.*
Roskill Information Services (1986). *The economics of carbon fibres.*
Shirley Institute (1986). *Blended fibre composites.*
Skeist Laboratories (1985). *Conductive plastics and plastics composites.*
Skeist Laboratories (1985). *Fire retardant plastics.*
SRI International (1984). *Titanium dioxide.*
Wolpert and Jones Studies (1984). *Catalysts in polymer production — an international report.*
Zinc Development Association (1983). *Zinc Oxide UV absorbers in plastics.*

4. Conferences

'*Carbon fibres*'. 3rd Conference, (8–10th Oct. 1985), London. Plastics and Rubber Institute.

'*Compounding for performance*'. (4–5th April 1984), London. Corporate Development Consultants.

'*Degradation of materials*'. (14–15th Oct. 1986), London. Instutute of Polymers.

'*Fillercon — fillers for plastics*'. (10–11th April 1984), Kingston, Surrey. Plastics and Rubber Institute.

'*Fillers '86*'. (13–14th Mar. 1986), London. Plastics and Rubber Institute.

'*Flame retardants*'. (Nov. 1983), London. Plastics and Rubber Institute.

'*Flame retardants*'. (28–29th Nov. 1985), London. Plastics and Rubber Institute.

'*Flame retardant chemicals — markets and trends*'. (4th Dec. 1986), London. Industrial Marketing Research Association.

'*Flammability and fire retardants*'. (7–8th June, 1984), Amsterdam. Alena Enterprises of Canada.

'*Flammability and fire retardants*'. (9–10th May 1985), Salzburg. Alena Enterprises of Canada.

'*Flammability and fire retardants*'. (20–21st June 1985), Montreal. Alena Enterprises of Canada.

'*Modifiers and additives*'. (6–7th Nov. 1985), New Jersey. Society of Plastics Engineers.

'*New technology to reduce fire losses and costs*'. (2–3rd Oct. 1986), Luxemburg. Queen Mary College, University of London.

'*Plastics additives — current markets and future trends*'. (1st Nov. 1984), London. Industrial Marketing Research Association.

'*Plastics compounding*'. (Sept. 1985), Birmingham. BIS Marketing Research.

'*Polymer additives, blends and composites*'. (10–11th April 1984), Luxemburg. Society of Plastics Engineers.

'*Polyolefines IV — innovations in processes, products, processing and additives*'. (27–28th February 1984), Texas. Society of Plastics Engineers.

'*Proper processing and product selection of flame retardants*'. (19–22nd Oct. 1986), U.S.A. The Fire Retardant Chemicals Association.

'*Silane and other coupling agents*'. (9th Oct. 1985), London.

CHAPTER ELEVEN

Properties of plastics and polymers

J. A. SHELTON

The spectrum of plastics and polymers now available, in both the thermoplastics and thermoset classes is vast, and this is being lengthened almost daily, not by entirely new polymer materials but by the 'blending or alloying' of like and unlike polymers to maximize the properties of two or more polymers in the synergistic sense. The previous chapter on 'Additives and catalysts' stressed the importance of fillers and reinforcements in transforming basic plastics into 'engineering plastics' with consequent potentials in structural applications outside the scope of the unfilled materials. The designer is therefore faced with a bewildering catalogue of materials. Since no two polymers are exactly alike in terms of properties or performance, selection is of critical importance, and publications such as the *Fulmer materials optimizer* (Fulmer Research Institute) are designed to assist in the selection for a given application.

We are undoubtedly living in the 'plastics age', and plastics offer design scope and flexibility, cost-effectiveness and functionality far beyond the capabilities of many alternative materials; indeed it is now possible for the polymer chemist to 'tailor' a polymer to a specific and exacting specification — the soft contact lens made from a hydrophilic acrylic polymer is a good example of such tailoring.

However, thermoplastics have their property limitations in view of their chemical structure, a long-chain carbon-based structure which allows them to be easily manipulated by heat and/or pressure. It also makes them susceptible to photo- and thermal-degradation, such degradation being more pronounced in certain polymer types than others. Furthermore, thermoplastics are subject to 'creep' which requires important consideration in structural end-uses.

Due criticism is often levelled at plastics by design engineers at the arbitrary way in which property data is presented by the materials manufacturers, which, unlike the traditional metals data for example, does not provide the engineer with a concise or clearly defined figure for use in his calculations. This 'scatter' in data presentation is due to the basic chemical structure of the polymer or polymer composite. It must be remembered that most thermoplastics have a 'softening point', a point at which some physical property becomes adversely affected, which is due to their amorphous state. Few thermoplastics have a clearly defined melting point, and in reality they soften gradually, going through a transition region so that the maximum safe design temperature will be well below the softening point. Conversely, plastics will become embrittled at low temperatures. The rheology of plastics is therefore a highly complex subject and their properties are strongly dependent upon time and temperature and sometimes humidity as well. A designer has to appreciate the classical or distinctive characteristics which are unique to plastics alone in order to interpret property data in a suitable manner. The complexity of these long-chain molecular structures makes it impossible to provide a complete quantitative prediction of properties from the present knowledge of chemical constitution and molecular configuration, but broad, qualitative correlations are possible.

The major properties of plastics can be categorized as follows: physical properties; thermal properties; mechanical properties; optical properties; electrical properties; chemical properties and weathering properties.

The physical properties are highly temperature-dependent. Most available data is based upon measurements made at 'ambient' temperature (20–23°C). Any deviation from ambient may result in significant changes in the tensile properties of most plastics, and will, in some cases, cause a deterioration in other properties. The thermosets are far less temperature-sensitive, but start to degrade chemically at certain temperatures.

Themoplastics behave as viscoelastic materials. Mention was earlier made of the 'creep' characteristics of thermoplastics which can be defined as the slow deformation of a material under constant stress, or the time-dependent strain which occurs when a material is subjected to a sustained stress. Few textbooks have been written on the subject, simply because creep and its characteristics were still under active investigation as late as the 1960s, Findley, Lai, and Onaran (1976) provided one of the first authoritative works on *Creep and relaxation of non-linear/viscoelastic materials*. It covered both metals and plastics, and described how the creep curves for many plastics (polymers) are similar to those for some metals. However, they usually do not exhibit a pronounced secondary stage. It also reviewed some of the empirical equations proposed to represent the creep curves of plastics.

The *Fulmer materials optimizer* (2nd edn) discusses the 'stress/strain' characteristics and creep characteristics of both metals and plastics, and makes mention of the fact that the viscoelastic creep behaviour of some thermoplastics also makes creep a possibility even at room temperature. Emphasis is placed upon this parameter since a knowledge of the stress-strain-time relationship is useful to designers in predicting the time-dependent deformation of a component which is subjected to a known system of stresses. Unlike most metallic materials, the creep of plastics is viscoelastic so that recovery (reversal of strain) to varying degrees takes place, even after the longest period of creep, once the load is removed.

The primary results of the creep test are usually presented as creep curves showing total strain against logarithmic time at different stress levels. The *Fulmer materials optimizer* discusses the creep testing of plastics in Vol 1, Part 1. Although British Standard BS4618 deals most comprehensively with the design of creep tests for plastics materials, there are as yet few data for plastics compared with those available for metals. Because of the scarcity of plastics creep data it is usually found necessary to circumvent the effects of creep by conservative design.

With regard to their thermal properties, plastics have high coefficients of thermal expansion — many times those of metals — and their dimensional stability is often dependent upon their previous 'history', so that when subjected to heat in any form, they may not expand uniformly in all directions. Ogorkiewicz (1970) considers the temperature dependence of thermoplastics and the need to understand the relationship between stress, strain, time, and temperature in their deformational behaviour. In view of the complexity of this relationship, it is advisable to confine attention to the effect of temperature in particular relationships between stress, strain and time.

The inherent coefficient of thermal expansion of plastics makes it essential to design 'in and for plastics', and not to attempt to copy or simulate an existing shape or configuration in, say, metal or cast iron; the PVC guttering and downpipes are a classic example of this rule. The early products, which simulated closely the cast-iron version, assumed a sinusoidal profile on very hot days, since they did not have the expansion facilities incorporated into modern systems. The expansion (and contraction) of these PVC gutters is plainly audible in hot weather.

A recurring problem associated with the properties of plastics is the difficulty experienced by design engineers in equating property data obtained from 'dumbbell' test pieces to their own specific needs — quite apart from the broad 'scatter' in such test figures. In consequence, many plastics components are over-designed to compensate for this uncertainty, particularly having regard for the variables that can influence the properties in the end-product such as a) orientation/flow in an injection mould, b) the cooling cycle, c) the absence of well radiussed corners at critical points in the design of the product, etc. For this reason, one or two references to mould design, and moulding processes are given in this chapter as essential reading in conjunction with those relating to the chemistry of polymers and their basic properties.

The relationship between properties and design in respect of the thermoplastics is well covered in Ogorkiewicz (1974). This is a collective work produced by ICI with the object of providing a general understanding of their properties, methods of characterizing them, and indicating how the data can be used in the design of thermoplastics articles. Of the various properties considered in the book, the mechanical ones are given most attention because they govern more frequently than others the more exacting applications of thermoplastics. Attention is further focussed on the type of information which is available or required, about the mechanical, as well as other, properties of thermoplastics, rather than the data themselves.

This collective work was produced by much the same team as that which prepared the earlier publication, Ogorkiewicz (1970), to which reference was made earlier in this chapter in relation to the temperature dependence of thermoplastics. It was originally intended to be based upon similar lines to this earlier one, but although a considerable amount of additional information had been produced, it was not sufficiently complete to warrant the preparation of another book centred on data. It was therefore decided to cover, in more general terms, progress made over the whole field of work on the properties of thermoplastics related to design.

It comprises nine, very readable chapters, the first two providing a broad background to the use of thermoplastics and a general introduction to their properties. The following six chapters describe the state of knowledge in 1974 of the different facets of the mechanical and other characteristics of thermoplastics, covering a) deformational behaviour, b) long-term durability, c) short-term strength and impact behaviour, and d) various factors affecting mechanical properties, electrical, thermal and various other physical and chemical properties as well as processing properties and methods. The last chapter describes the use of mechanical design data and presents a number of case studies.

It contains some 63 references relating to a diversity of subjects such as 'creep' of glassy polymers; predicting end-product performance; designing with thermoplastics in blending; sandwich moulding; fatigue; design principles for calculating the size of rib reinforced sheets of thermoplastics, etc. The index is adequate.

The previous chapter on 'Additives and catalysts' refers to the role of fillers and reinforcements in upgrading the properties of plastics throughout the polymer spectrum, and the important contribution that fillers have made in enhancing the properties of plastics (especially the thermoplastics), cannot be over-emphasized. Fillers and reinforcements have always played a major part in functionalizing certain thermosets, which, in their unfilled form are too brittle to be of any commercial value.

It is in the field of thermoplastics that fillers, and in particular the mineral fillers, have played a major part in property enhancement to the extent that lower cost so-called commodity plastics can be transformed into highly functional materials, capable of being used in demanding 'engineering' or 'structural' applications.

Fillers will enhance a wide range of properties within the base polymer matrix, but they also affect the processing behaviour. The fillers used can be categorized as inorganic and organic, the inorganics being further subdivided into non-fibrous and fibrous. Examples of non-fibrous inorganics include talc, calcium carbonate, mica, etc. Fibrous fillers include glass, metallic, and ceramic fibres, and the organics include materials such as sisal, wood flour, etc.

The fillers/reinforcements are sometimes used in hybrid form, such as mixed particulate or fibre, or mixed particulate/fibre systems.

The properties influenced by the filler systems can be subdivided into 1) end use, which will include static and dynamic stiffness and strength (including impact, creep and fatigue); wear resistance; density; high and low temperature resistance; thermal conductivity; electrical conductivity; electrical breakdown resistance; chemical resistance; weathering; fire retardancy; radiation shielding, and 2) processing, including shrinkage and warpage; rheology (e.g. viscosity, flow ratio); melt abrasivity; colour; and surface finish.

With the advent of structural foams and reaction injection mouldings, it is essential to avoid any confusion between 'stiffness' and 'strength', since these low density mouldings will exhibit high stiffness (in both the unfilled and filled form) but will lack the strength of a solid moulding.

Reverting again to the role of additives, a greater understanding of how they affect specific properties is beginning to emerge, but nevertheless, more detailed models of the real structure would allow a quantitative prediction of properties.

The 'hybrid' filler systems provide scope for 'tailoring' composites to specific applications.

In addition to improved theoretical prediction methods, it is desirable to have quicker and more informative test methods to evaluate sample sheets or actual products. The weaknesses of long established test methods are shown up when testing the inhomogenous and frequently anisotropic samples associated with high filler concentrations. The effects on impact behaviour are not revealed in sufficient detail by Izod or Charpy methods. Unidirectional quasi-static tests require supplements with data from more complex loading. It is interesting to note that researchers into filled thermoplastics have readily adopted instrumented impact testing methods, frequently concentrating on plaque rather than bar tests. The information provided by such test methods is much greater than that from non-instrumented conventional impact tests and substantial effort may be required to fully understand the impact behaviour, but the eventual insight provides substantial advantages in developing improved materials.

In a review of fibre reinforced plastics Judd (1983) describes their favourable specific strength and modulus properties. The fibre can, in many cases, be oriented in the direction of principal stress. The mechanical properties depend upon 1) the reinforcing fibre, its aspect ratio and its surface treatment, 2) the orientation of the fibre as a tow or in a cloth, and the type of weave, and 3) the resin/reinforcement ratio; the mechanical properties of the composite being proportional to the volume fraction of the fibres.

Included among the matrix resins for these composites are phenolics, furans, unsaturated polyesters, epoxies, polyimides and certain thermoplastics. Jute, glass, synthetic textile fibres, and the higher modulus Kevlar, boron and carbon fibres are used as the reinforcement in a variety of forms including continuous filaments, cloth, chopped fibre and chopped fibre mat.

The review gives comparative property data between these fibre reinforced plastics and metals in terms of stress-strain behaviour, mechanical properties, fatigue resistance, etc. It also gives a cost comparison of the reinforcing fibres based upon 1983 prices.

The synergistic link between properties and good design and vice versa cannot be over-emphasized when applying plastics. Certain publications on this subject, although listed at the end of this chapter, have a somewhat limited value in guiding the uninitiated; for example, Levy, S. and Dubois, J. M. (1984) which is the 2nd edn of a plastics product design engineering handbook, contains a number of errors and misleading or vague statements. In Florscheim, C. H. (1983) the chapters on material selection, particularly those dealing with plastics and rubbers, give little more than a cursory introduction to the subject, and only serve to emphasize the fact that 'a little knowledge is a dangerous thing'. Such is the complexity of the polymer industry and its associated materials and processes (without including the rubbers and elastomers) that any book which endeavours to cover the subject in any form other than within a very narrow or specific subject spectrum, must inevitably contain a degree of generalisation. This is true in the case of Powell (1983) which is designed to educate students in the art of engineering with polymers. It deals with four basic aspects: 1) terminology and general properties of plastics, 2) stiffness, strength and fracture behaviour on these properties, 3) property modification, e.g. by fibre reinforcement etc., and 4) thermal and flow properties of polymer melts. These comments are not made by way of criticism of the author, for this book fulfils its intention. Until plastics become a major subject in our education system, the author of any publication on their properties will be faced with the impossible task of satisfying a readership whose knowledge of polymers many range from 'what is a polymer?' to the polymer chemist requiring the latest thesis on a specific parameter.

The proceedings of The Society of Plastics Engineers and those of Society of the Plastics Industry in the USA published annually, make a particularly valuable contribution to the furtherance of knowledge in respect of properties since they contain papers on very specific subjects, grouped under respective divisional headings, i.e. *Engineering Properties and Structure Division*, etc. These papers have a high technical content, presented in a more or less standard format, in a very readable form.

The *Designers guide to materials* by the British National Committee on Materials comprises a number of individual publications in report form on all facets of materials properties and uses. The property information section (Section 3), covers metals, plastics, elastomers, composites, and adhesives. Section 4 deals with design information in respect of strength calculations, selection of materials, both metallic and non-metallic, and design considerations. One contribution by ICI Petrochemicals and

Plastics Division deals with the presentation and use of data on the mechanical properties of thermoplastics, with 15 references which, although somewhat dated (the original paper was prepared in 1973) is particularly pertinent to this chapter on properties.

A further contribution, from an unspecified source, provides a guide to the selection of plastics materials in tabulated form. It makes the valid point that 'for only a few applications will there be an ideal material. The correct selection will normally only lead to the best compromise resulting from different requirements', and in order to make the optimal choice, it is most important to know exactly what is required of the part to be played by the plastics material. It is therefore advisable that the designer first makes a list of requirements for the part, and then applies a weighting to the various requirements in order to arrive at the best choice of material.

References

1. Textbooks

Allen, N. S. (1983). *Degradation and stabilisation of polyolefines* (Elsevier).
Andrews, E. H. (1979). *Developments in polymer fracture* (Applied Science).
Bhattacharya, S. K. (1986). *Metal-filled polymers — properties and applications* (Marcel Dekker).
Brown, R. P. (1981). *Handbook of plastics test methods* (George Godwin).
Brydson, J. A. (1981). *Flow properties of polymer melts* (George Godwin).
Brydson, J. A. (1982). *Plastics materials* (4th edn, Butterworths).
Burns, R. (1983). *Polyester moulding compounds* (Marcel Dekker).
Carraher, C. E. and Seymour, R. B. (1984). *Structure-property relationships in polymers* (Plenum Press).
Cheremisinoff, N. P. and Cheremisinoff, P. N. (1978). *Fiberglass — reinforced plastics deskbook* (Ann Arbor Science Publishers).
Cogswell, F. N. (1981). *Polymer melt rheology* (George Godwin).
Davis, A. and Sims, D. (1983). *Weathering of polymers* (Applied Science).
Deanin, R. D. (1972). *Polymer structure, properties and applications* (Cahners).
Deanin, R. D. and Crugnola, A. M. (1976). *Toughness and brittleness of plastics* (American Chemical Society).
De Dani, A. (1960). *Glass fibre reinforced plastics*. (Newnes).
Ehrenstein, G. W. and Erhard, G. (1984). *Designing with plastics* (Hanser).
Findley, W. N., Lai, J. S. and Onaran, K. (1976). *Creep and relaxation of non-linear viscoelastic materials* (North-Holland).
Florscheim, C. H. (1983). *Industrial design in engineering, a marriage of techniques* (The Design Council).
Folkes, M. J. (1983). *Short fibre reinforced thermoplastics* (Wiley).
Fredrickson, A. G. (1964). *Principles and applications of rheology* (Prentice-Hall).
Garner, D. P. and Stahl, G. A. (1983). *Effects of hostile environments on coatings and plastics* (American Chemical Society).
Gould, R. F. (1968). *Stabilisation of polymers and stabilizer processes* (American Chemical Society).

Grassie, N. (1985). *Development in polymer degradation*, Vol. 1–6 (Elsevier).
Grassie, N. and Scott, G. (1985). *Polymer degradation and stabilisation* (Cambridge University Press).
Harper, C. A. (1975). *Handbook of plastics and elastomers* (McGraw-Hill).
Harris, B. (1983). *Developments in GRP technology*, Vol. 1 (Elsevier).
Hawkins, W. L. (1984). *Polymer degradation and stabilisation: properties and applications*, Vol. 8 (Springer-Verlag).
Hepburn, C. and Reynolds, R. J. W. (1979). *Elastomers: criteria for engineering design* (Applied Science).
Hilyard, N. (1982). *Mechanics of cellular plastics* (Applied Science).
Holister, G. S. and Thomas, C. (1986). *Fibre reinforced materials* (Elsevier).
Jellinek, H. (1983). *Degradation and stabilisation of polymers*, Vol. 1 (Elsevier).
Katz, H. S. and Milewski, J. V. (1978). *Handbook of fillers and reinforcements for plastics* (Van Nostrand Reinhold).
Kausch, H. H. (1978). *Polymer fracture* (Springer-Verlag).
Kinloch, A. J. and Young, R. J. (1983). *Fracture behaviour of polymers*. (Applied Science).
Lenk, R. S. (1978). *Polymer rheology* (Applied Science).
Levy, S. and Dubois, J. H. (1984). *Plastics product design engineering handbook* 2nd Edn, (Chapman and Hall).
Lubin, G. (1969). *Handbook of fiberglass and advanced plastics composites* (Van Nostrand Reinhold).
Manson, J. A. and Sperling, L. H. (1976). *Polymer blends and composites* (Plenum Press).
Margolis, J. M. (1985). *Engineering thermoplastics properties and applications* (Marcel Dekker).
McKelvey, J. M. (1962). *Polymer processing* (Wiley).
Meares, P. (1965). *Polymers: structure and bulk properties* (Van Nostrand Rheinhold).
Miller, M. L. (1966). *The structure of polymers* (Van Nostrand Reinhold).
Nielson, L. E. (1974). *Mechanical properties of polymers and composites* (Marcel Dekker).
Neiman, M. B. (1965). *Aging and Stabilisation of polymers* (Consultants Bureau).
Ogorkiewicz, R. M. (1970). *Engineering properties of thermoplastics* (Wiley-Interscience).
Ogorkiewicz, R. M. (1974). *Thermoplastics properties and design* (Wiley).
Owen, E. D. (1984). *Degradation and stabilisation of PVC* (Elsevier).
Pinner, S. H. (1966). *Weathering and degradation of plastics* (Columbine).
Powell, P. C. (1983). *Engineering with polymers*. (Chapman and Hall).
Pritchard, G. (1984). *Developments in reinforced plastics*, Vol. 3 (Elsevier).
Pritchard, G. (1984). *Developments in reinforced plastics*, Vol. 4 (Elsevier).
Saechtling, H. (1983). *International plastics handbook* (Hanser).
Schnabel, W. (1981). *Polymer degradation* (Hanser).
Sheldon, R. P. (1982). *Composite polymeric materials*. (Elsevier).
Shlyapintokh, V. (1984). *Photochemical conversion and stabilisation of polymers* (Hanser).
Stepek, J. and Daoust, H. (1983). *Polymer properties and application — additives for plastics* (Springer-Verlag).
Struik. L. C. E. (19780. *Physical aging in amorphous polymers and other materials* (Elsevier).
Tadmor, Z. and Gogos, C. G. (1979). *Principles of polymer processing* (Wiley).
Titow, W. V. and Lanham, B. J. (1975). *Reinforced thermoplastics* (Applied Science).
Titow, W. V. (1984). *PVC technology* 4th Edn, (Elsevier).
Throne, J. L. (1979). *Plastics process engineering* (Marcel Dekker).

Troitzsch, J. (1983) *International plastics flammability handbook* (Hanser).
Ward, I. M. (1971). *Mechanical properties on solid polymers* (Wiley).
Ward, I. M. (1975). *Structure and properties of oriented polymers* (Applied Science).
Williams, J. G. (1980). *Stress analysis of polymers* (Ellis Horwood).
Wyatt, O. H. and Dew-Hughes, D. (1974). *Metals, ceramics and polymers* (Cambridge).
Yescombe, E. R. (1976). *Plastics and rubber — world sources of information* (Applied Science).

2. Publications

Additives for Polymers (Monthly) (Elsevier/Fulmer Yarsley).
ANTEC Proceedings (Annually) (Society of Plastics Engineers).
ANTEC Proceedings (Annually) (Society of the Plastics Industry)
Appraisal of Weathering Behaviour of Plastics (1973) (Building Research Establishment).
Encylopedia of Composite Materials and Components (Technomic Publishing).
Encylopedia of Polymer Science and Engineering, 2nd Edn (1985) (Wiley).
Engineering Design Properties of GRP (1979) (British Plastics Federation).
European Plastics Buyers Guide (Annually) (IPC Industrial Press).
Fatigue and Creep in Reinforced Plastics (1967) (British Plastics Federation).
Fulmer Materials Optimizer (1974) (Fulmer Research Institute).
International Plastics Selector(1977) (Cordura).
Materials Selector and Design Guide (1974) (Morgan-Grampian).
Modern Plastics Enclyclopedia (Annually) (McGraw-Hill).
Plastics, Edition 8 (1986) (DATA).
Plastics Industry Directory (1984) (Maclaren Publishers/British Plastics Federation).
Plastics Materials and Processes (1982) (Van Nostrand Reinhold).
Plastics World Directory (Annually) (Cahners).
Recommendations for Presentation of Plastics Design Data — General Introduction (British Standard B.S.4618:1970).
Users Practical Selection Handbook for Optimum Plastics, Rubbers, and Adhesives (1976) (International Technical Information Institute, Japan).

3. Study Reports

Battelle (1984). *Polymer blends.*
Battelle (1986). *Carbon fibre reinforced polyphenylene sulphide.*
Business Communications (1984). *Plastics fillers and extenders.*
Business Communications (1984). *Plastics UV stability.*
Industrial Aids (1986). *Thermoplastic composites.*
Kline (1985). *Advanced polymer composites.*
Kossoff and Associates (1984). *Compounding.*
Phillip Townsend Associates (1987). *Thermoplastics compounding.*
Shirley Institute (1986). *Blended fibre composites.*
Skeist Laboratories (1985). *Conductive plastics and plastics composites.*

4. Conferences

'*Compounding for performance*'. (4–5th April 1984), London. Corporate Development Consultants.
'*Degradation of materials*'. (14–15th Oct. 1986), London. Institute of Polymers.

'*Modifiers and additives*'. (6–7th Nov. 1985), New Jersey. Society of Plastics Engineers.

'*Plastics compounding*'. (Sept. 1985), Birmingham. BIS Marketing Research.

'*Polymer additives, blends and composites*'. (10–11th April 1984), Luxemburg. Society of Plastics Engineers.

'*Polyolefines IV — innovations in processes, products, processing and additives*'. (27–28th February 1984), Texas. Society of Plastics Engineers.

CHAPTER TWELVE

Business information

R. T. ADKINS

Making polymers and plastics products, but selling their properties is a recognizable feature of the industry, and the struggle of each of the manifold varieties, modifications, grades and forms of plastics materials and their applications to find their niche in the market place is a highly information dependent process. Making a profit from a product which may have taken hundreds of man-years to conceive, manufacture and market in the face of intense competition from other manufacturers of similar or alternative products demands a high degree of business awareness. Competitive edge is gained from having the best information at the time when it is most wanted, usually instantaneously and urgently, 'yesterday' being a frequent response to asking for an indication of how soon the information might be required.

In common with purely technical information, there is no one universal source of business information, but in many respects the types of publication and service catering for the one are to be found paralleled in the other sector. For example, like their colleagues in laboratory, plant and workshop, business people need reliable numerical data which may be subjected to statistical manipulation, which is reflected in the continuing growth of computer-based services supplying this need.

No less important are other factual data concerning company and government activities at home and abroad, names, products, ownership, trends, cycles and forecasts.

Providing useful business information for decision making is not just a question of finding suitable publications and reproducing the contents, but of adding value, by marshalling the relevant information into a coherent and credible whole from which dependable conclusions may be reached.

Smith (1985) has indicated a successful procedure based on many years practical experience of the needs of a demanding group of users in the chemical industry. Although the collection of data and information is an essential part of the preparation of the 'intelligence' report, without the evaluation which only a thorough understanding of the problem which motivated the search can reveal, its significance may be lost.

This guide however can only conveniently discuss the existence and content of business information sources chiefly with regard to the medium and format in which they are presented, in the hope that their potential for producing the essential added value, 'intelligence', will be adequately signposted. These sources range over such a wide variety of business activities (including those not specifically concerned with plastics per se, such as financial services and general management) that comprehensive coverage of everything which may conceivably be of value is not possible.

Dr Samuel Johnson's often quoted, somewhat facile dictum that 'we either know a subject or know where information on it may be found' is true only insofar as the information is structured to allow it to be examined in a systematic way. This guide will at least provide some signposts to the different media used and their, often different, treatment of the same information.

Directories

There are a number of general business directories available in many of the larger libraries which suggest useful, obvious starting points for searching, often the most difficult part of the search. G. Smith, in *Business information sourcebook* (Headland Press 1985), a frequently updated looseleaf compilation, gives some useful insights into the nature of business information with good coverage and discussion of directories, as well as newspapers, newsletters, journals, looseleaf and card services, reports, statistics and monographs.

Company information sourcebook, a companion work by Headland Press evaluates the major UK company sources in considerable depth and is updated twice yearly.

The *Macmillan directory of business information sources* (J Tudor: Macmillan Publishers) (1987) discusses sources and information centres, broadly classified by the Standard Industrial Code (SIC) which includes the headings:

2514 Synthetic resins and plastics materials
2515 Synthetic rubber

3275 Machinery for working wood, rubber, plastics, leather and paper
4811 Rubber tyres and inner tubes
4812 Rubber products
4831 Plastics coated textile fabric
4833 Plastics coated floor coverings
4834 Plastics coated building products
4835 Plastics packaging
4836 Plastics products not elsewhere specified.

Under the last heading are to be found sources such as the Zip Fastener Manufacturers Association, the Furniture Industry Research Association and the Horticultural Trades Association. SIC hierarchical codes are maintained by the UK Central Statistical Office (CSO) and are commonly used for the classification of company products and services.

Another general, comprehensive directory is the *International directory of business information sources and services (IDBISS)* published by Europe Publications Ltd., arranged by country, and within country by organisation. These include chambers of commerce, foreign trade promoting organisations, government-, independent- and research organisations, sources of statistical information and business libraries. These contacts listed are often only the starting points for the search, but nevertheless offer useful reminders to the many different possible avenues of approach. Some organisations are more helpful than others, often a response being dependent on membership or increasingly, willingness to pay a charge for the service.

Current British directories (CBD Research Ltd) is a title index to the directories of British Trade Associations and other regional, telephone, city and town directories, indexed by subject and publisher.

Handbook of international trade provides a useful review of published resources covering sources of information on exporting, standard reference works, directories, statistics, marketing data and periodicals.

Earlier material is well covered by Yescombe in *Plastics and rubber: world sources of information* (Applied Science 1976).

Other 'old faithfuls', those familiar general directories leading directly to information on companies active in the plastics and related industrial sectors appear below.

Guide to Key British Enterprises (Dun and Bradstreet Ltd.) is published in colour-coded sections listing 20,000 company names, products and trades under SIC codes and geographical area breakdowns.

Kompass (Kompass Publishers, a Division of Information Services Ltd.) is a much more comprehensive publication, with individual country volumes containing tables of coded company activities and products arranged as a 'dot-matrix' tabulation from which the company name can be retrieved immediately, which may be followed up with the full company product and financial data from alphabetical listings (arranged by town and country) in other volumes. In-depth indexing of products and services gives good discrimination between form, materials, processes and uses of plastics materials. A helpful feature in the European volumes is an English index equivalent to the native language index.

Sell's Directory (Sell's Publications) currently on its 102nd edition, covers more than 25,000 products and services listed alphabetically with uniquely coded cross-references to more than 60,000 companies and 10,000 trade names. Subject breakdown, as in so many similar directories is not exhaustively detailed, many headings, for instance, plastic dustbins, referring to but one manufacturer, while under others, e.g. plastic goods or plastics fabricators, many code numbers are listed. This does not necessarily detract from its usefulness, but is an indication of the compromises which have to be made when compiling such comprehensive works.

Kellys Directory (Kellys Directories) of equal longevity, is similar in many other ways to Sell's, offering broadly classified trades and services, company information and trade names but with additional information on UK exporting and overseas companies.

Thomas register of American manufacturers and catalog file is also well established, the 75th edition appearing in 1985. Published in 19 volumes, volumes 1 to 11 contain alphabetical lists of products and services, by state and city within the state. Volumes 12 and 13 contain company details and trade names, and volumes 14 to 19 (*Thomcat*) contain reproductions of company catalogues.

The Science Reference and Information Service (SRIS) of the British Library at Holborn, London has an extensive collection of directories in the Business Information Section and publishes *Business information*, a brief, classified guide to their holdings. This publication also contains commentaries on much else of interest and value on companies, products, market and statistical sources which is on open access and may be consulted immediately.

National directories of the plastics industry in countries throughout the world have a number of features which are fairly consistently found, e.g. an alphabetical listing of company names and other details, a section classified by product, process, equipment, form, service, use, etc., a trade name index and sometimes a report of the general business activity in each sector accompanied by statistical data on imports, exports and production. Some may be multilingual, but often include an English language version.

Totally comprehensive trade directories have yet to be produced and they sometimes leave serious gaps in their coverage. Nevertheless as broadly classified sources they have a useful role, often providing data quickly and acting as useful starting points for gathering information on the main companies in specific sectors. Some of the major directories and their main features appear below.

Australian plastics and rubber buyers' guide (IPC Business Press (Australia) Pty. and Plastics Institute of Australia). Company names and addresses are classified by raw materials, products, processes, machinery, trade names, agencies and indexed by subject.

Buyers' guide to plastics and chemical products (Kompass Publishers, East Grinstead and the British Plastics Federation (BPF)) (1986). Based on the UK *Kompass directory*, this is a handy guide to over 6000 different plastics, rubber and chemical products. Over 5000 companies, including BPF members are listed.

British Plastics Federation buyers' guides are published as a series of individual, compact 'dot-matrix' compilations of company activities in, for instance, plastics packaging, moulded products and services, reinforced plastics chemical plant, plastics additives, rotational moulders, thermoplastics and thermosetting materials, thermoplastics pipes and fittings, plastics and rubber processing machinery and equipment, and reinforced plastics.

Canadian plastics directory and buyers' guide (Southam Business Publications) has indexes to processors, products and suppliers of polymers, additives, etc., with separate alphabetical company listings. An index to trade names, resins, compounds and additives is provided and machinery manufacturers, mould, tool and die makers are listed.

Chemical industry directory and who's who (Benn Brothers Information Services, Tonbridge, annual). Chemical manufacturers and trader details are presented with suppliers of laboratory equipment and chemical plant. Lists of trade names, research associations, professional and trade organisations and independent consultants are given.

Die kunststoff industrie der Schweiz (Verlag fur Wirtschaftliteratur) is produced in both French and German with alphabetical listing by company and a classified product section.

Die kunststoff industrie und ihre helfer (Industrieschau Verlag) is index in German, English, French and Italian, with companies listed alphabetically within individual towns.

European plastics buyers' guide (IPC Industrial Press, London) contains English, French, German, Italian and Spanish indexes to products and services in serial number order (not SIC) with listings of companies and trade names.

France plastiques — annuaire officiel des plastiques (CEPP Creations Editions Productions Publicitaires, Paris) is classified by primary materials equipment, services, fabrication process, and includes listings of trademarks and companies.

Guide to rubber and plastics test equipment. 2nd edn. ed. R.P. Brown (RAPRA Technology Ltd., Shawbury, 1979). Useful both as directory and reference text to data this is compiled and assessed by RAPRA from suppliers' test equipment. Chemical analysis is included and a product index identifies suppliers.

Guide du syndicat général des commerce et industrie du caoutchouc et des plastiques (SGCICP, Paris), is classified by product, company, and region of France with indexes to brands, trade names and suppliers. There are sections on material properties, production statistics and import and export figures.

Gummi addressbuch mit bezugsquellennachweis (Curt R Vincentz Verlage, Hanover) furnishes information on companies, organizations, institutions listed alphabetically by town and classified by product, machinery and service.

Hong Kong plastics (Wah Sun Hong and Star Industrial) The Hong Kong plastics industry, markets, exports and manufacturers are surveyed with classified listings of manufacturers by product.

Industry directory (British Plastics Federation and Kompass Publishers) have produced a comprehensive guide to UK moulding, converting and fabricating services, classified and arranged in the well-known Kompass 'dot-matrix' format by company product and service.

Kunststoffe plastics (Vogt-Schild AG Druck und Verlag, Solothurn) gives French and German SIC indexes to trade associations, and lists companies within the Swiss plastics industry.

L'officiel des plastiques et du caoutchouc (Paris) contains lists of trade names and discusses imports, exports, consumption and properties of materials, fabricators and importers of main classes of materials and products, mould makers for plastics and rubber moulders, and an unusual feature, standards in force.

One of the most important annuals in the English language in the plastics industry, especially for USA-oriented information, *Modern plastics encyclopaedia* (McGraw-Hill) is published in four sections, (1) a Textbook section discussing materials, chemicals, additives, fillers, property enhancers and reinforcements, primary processing, including auxiliary, tooling and testing techniques, fabricating and finishing; (2) a Design Guide devoted to the selection of the appropriate plastic with the right performance for the application; (3) a Data Bank containing property, design, chemical, additive and machinery data with (4) a subject classified list of 5000 suppliers and an alphabetical company name index.

New trade names in the rubber and plastics industries (RAPRA Technology, Shawbury, Annual). This is a useful compilation for tracking down elusive trade names which have been described in the literature. The coverage is wide, including thermoplastics, thermosets, rubbers, additives, processing aids, compounds, machinery and test equipment, finished and semi-finished products. For each entry a description is given of the product, company name and details of the first reference to the product. Company addresses are listed and the trade names are also classified by product category, a useful buyers guide supplement.

OPD chemical buyers' directory — chemical marketing reporter, (Schnell Publishing, Annual). Known as the 'Green Book', over 800 chemicals suppliers, and their telephone numbers are listed in a quick reference format. Company catalogues and brochures are also included.

Plasticos españoles (Spanish National Association of Plastics Industries — ANAIP) is sponsored by the Directorate General for Exports of the Ministry of Economy and Commerce. Basic facts and figures on the Spanish industry with company lists and a product index are given.

Plexconcil silver jubilee directory (Plexconcil — Plastics and Linoleum Export Council, Eastra, Bombay). Production and use figures are given for a number of plastics, current members are listed and companies are classified by somewhat heterogenous sectors comprising blow moulders, cables, machinery, masterbatch, raw materials and toys.

Rubber and plastics news rubber directory and buyers' guide — Rubbicana (North American Rubber Producers, Manufacturers and Rubber Industry Suppliers, Crain Automotive Group). Company names are listed, classified by product, services, machinery and, an unusual feature in this type of publication, a who's who of personalities in the trade is provided.

Rubber red book (Communication Channels) published in the journal *Elastomerics*, contains lists of manufacturers and suppliers of products and services in the USA, Canada and Puerto Rico with subject indexes to the classified lists.

Rubber blue book — materials, compounding ingredients and machinery for rubber (Lippincott and Peto, Annual). Six main sections cover rubber and latex compounding materials, their trade names, properties, compounding, processing, colouring, vulcanisate applications and properties.

Rubbicana Europe (Crain Communications, London, 1987) lists over 1000 suppliers of materials and equipment and over 1300 manufacturers of rubber and polyurethane based products.

Skandinavisk plast-industri (Scandinavian plastics industry) (Okonomisk Literatur Norge), covers Denmark, Norway, Sweden and briefly, Finland, with English summaries, with classified country, company and product information.

SPI directory and buyers' guide (Society of the Plastics Industry, New York, Annual), gives regional listings of all SPI member companies, products and services. Individual members' expertise is included.

The synthetic rubber manual (formerly the *Elastomers manual*) (The International Institute of Synthetic Rubber Producers). Major rubber types, classified by the ISO classification system, produced by European, Far Eastern, Latin American and North American manufacturers are listed.

UK directory of injection moulders of thermoplastics (Corporate Development Consultants, Bristol) lists over 1300 injection moulding companies in alphabetical order and by 11 UK geographical regions.

US foamed plastics markets and directory (Technomic Publishing) discusses foam market trends and classifies companies by product.

Yearbook of the Italian rubber industry (Gesto Srl.) contains a geographical listing of Italian rubber manufacturers, an alphabetical listing of Italian rubber article manufacturers, manufacturers classified by product, suppliers of raw materials, machinery and services, lists of companies and their agents and tradenames.

Trade directory information in journals

Many journals carry an annual compilation, although some listings feature in each issue, e.g. the Buyers' Directory in *British Plastics*. These are timely and convenient supplements to the hard-backed annual versions, particularly useful because they reflect changes more rapidly.

Plastics and rubber sector journals include the *Plastics Technology Manufacturing Handbook and Buyers' Guide* (Bill Communications, Annual). Published as an additional issue of *Plastics Technology*, the directory covers injection moulding, extrusion, blow moulding, thermoforming, urethane systems, RIM, calendaring, structural foam, reinforced plastics equipment, rotational moulding, compression transfer, decoration, printing, finishing, size reduction, welding, sealing, compounding, mixing and blending, chemicals additives and plastics materials.

Other journals include *British Plastics and Rubber*, the January issue of which is devoted entirely to trade information, *Canadian Plastics, Kunststoff Journal, Kunststoffe Berater, Kunststoffe, Plastforum Scandinavia, Plastics World, Rubber World.*

Classified trade information may also be found in *Angewandte Chemie, Chemiker Zeitung, Converting World, Farbe und Lack* and *Ingenieria Quimica.*

Abstracts journals

Some of these sources have been covered in other sections, with different emphases, here the focus will be on the business information content.

RAPRA Abstracts covering 400 periodicals, conference proceedings, books, trade and technical literature is published monthly by Pergamon Press, with annual subject, author and company name indexes appearing after a considerable delay in the following year. The indexing is reasonably specific with main subject headings e.g. Company Information, subdivided by company name, polymer, application, process, property, etc. Marketing is also broken down into country, polymer and use etc. Each issue contains a Commercial and Economic Section subdivided by company activities, production, organization and planning, purchasing and costing, industrial relations and personnel, marketing, and statistics and economic information, which is also further categorized into general, machinery, raw materials and monomers, polymers, compounding ingredients, polymeric forms, powders and latices, semi-finished products, fibre laminates, finished products and applications.

Trends in End-Use Markets for Plastics is published monthly by Springborn Laboratories (N.America) and the contents are classified by specific market, from adhesives, agriculture through to sports with detailed abstracts and tables. Price ranges of monomers, polymers and bulk chemicals for Europe and the USA are included.

Urethane Abstracts is published monthly by Technomic Publishing Co., Philadelphia, as a current awareness bulletin containing market and company information, personnel movements and plant capacity data.

Customized retrospective searches from these files may be undertaken by the Science Information Services Department of the Franklin Institute Research Laboratories.

Business International Index is published as a weekly newsletter by Business International Corporation, London, giving details of monthly, bimonthly and annual results for company activities in the plastics industry in East and West Europe, Asia, China and Japan with cumulated quarterly and annual indexes based on an extensive thesaurus of free text terms.

PROMT (Predicasts Overviews of Marketing and Technology, Monthly), is one of several important data bases produced by Predicasts Inc., Cleveland, which, like RAPRA, are available in both printed and online form. *PROMT* contains abstracts of significant articles on companies, products, their markets and technology appearing in over 1000 periodicals, newspapers and reports worldwide. These are indexed by SIC product codes in the cumulated cross-referenced, general subject index which gives the Section, abstract number and issue for each abstract.

Other monthly publications are the *Predicasts F & S International, Europe and United States Indexes*, which contain brief but informative notes on companies, industries and products. *F & S International* covers Latin America, Mid-East, other Asia, Canada, Africa and Japan; Europe covers the EEC, Scandinavia, other West Europe, USSR and other East Europe. The quarterly Source Directory lists titles, supplemented by country and subject lists, e.g. under Chemicals and Allied Products. Alled Products is a catch-all term including many relevant terms useful for the searcher, e.g. over 185 titles on Polymers, Plastics, Rubbers and Fibres are listed.

World Product Casts (Quarterly) contains regularly updated forecasts packaged individually as country and product files, e.g. P-3 contains information on Chemicals, Polymers, Drugs, Oil, Rubber, Stone, Clay and Glass arranged by SIC code, Plastics 2821000 subsuming for instance Plastics, Polyolefin 2821400. Products are tabulated in SIC code order by Product, Event (a kind of aspect or role indicator, e.g. consumption), country, year range, quantities, units, source reference and annual growth rate.

Chemical Industry Notes (Weekly, Chemical Abstracts Services) covers over 80 business and trade journals and concentrates on business, financial and marketing activities within the chemical industry. An annual volume index is produced.

Chemical Business Update (Royal Society of Chemistry, Monthly), usefully covers the 'grey literature' appearing in the online data base *Chemical Business Newsbase*. 'Grey literature' is defined as literature other than serials available over thc counter and includes 'stories from company reports, news releases, newsletters, market research and stockbroker reports, directories and the like'. Abstracts include numerical data and three indexes, subject, chemical product and company and corporate source are provided.

The highly condensed, monthly *Business Periodicals Index* (H.W. Wilson Co.) cumulated quarterly and annually, is a counterpart to the well established *Technology Index* by the same firm, which publication itself should not be ignored as it covers such topics as marketing, advertizing, market research, market statistics, surveys, export trade, financial risk, commercial law and policy.

Business Periodicals Index, a useful source which is very rapidly scanned, covers a wide range of trade, financial and management journals, listing the article titles and minimal, but adequate journal reference under highly specific, controlled subject headings with efficient cross-references.

Science Citation Index and Current Contents (Institution for Scientific Information — ISI) cover a wide range of journals of relevance to the industry. Although perhaps not oriented specifically to the extraction of commercial information the indexes are easily scanned.

Periodicals

Some have already been mentioned in the section on directories but another important function is to provide up-to-date information and news on individual company activities, product and process developments and large-scale industry reviews. There are many such journals, ranging from the specialised to the more general, often reflecting national concerns, often providing the first and sometimes the only source, of information ranging in value from the ephemeral to the substantial and permanent. Equally it must be said that there is much duplication of information across a range of journals, although the level of detail, commentary and interpretation will vary from journal to journal.

Tattum (1987) suggests in an interesting article on the history and development of *European Chemical News* (ECN) over the past 25 years, 'that the trade press can be seen as a mirror, reflecting significant events across a particular area of business', a theme illustrated by some of the issues reported on during that time. Journals such as *ECN* however, are genuine opinion formers as well as reflectors of current events and because their influence is far reaching, they are important reading for the executive, and a rich source for retrospective searching. Much information on plastics materials is to be found not only in the specialist journals covering the industry but as with *ECN*, appearing in the international chemical industry coverage, and because of their importance highlighted in monthly monitors.

A somewhat subjective selection therefore of the large number of journals containing potentially useful business information is given below. Space precludes detailed analysis of features, but most journals run true to type with sections on companies' activities in response to market changes, expansion, retrenchment, investment, regrouping, economic trends and outlook, statistical data, government and legislative announcements, editorials, products and applications.

GENERAL CHEMICAL PERIODICALS

Chemical Age (International, weekly) gives useful consumption figures for the industry.

Chemical Business (monthly, Schnell Publishing) discusses trends in the international chemical marketplace, new products and applications mainly with reference to the USA.

Chemical and Engineering News (American Chemical Society, weekly) gives quarterly earnings for the major international chemical companies.

Chemical Industries Newsletter is published six times per year by the Process Industries Division of SRI International covering current research undertaken by the Petrochemicals, Polymers and Energy Center, among several others, on technoeconomics, marketing, diversification, price forecasting and business opportunity evaluation.

Chemical Insight is compiled and published fortnightly by Mike Hyde as a perspective on the international chemical industry. The comments are crisp, the statistics digestible, and the style faithfully reflects the business ethos of the management of the major international organizations in a very readable way for which Mike Hyde has gained a high reputation.

Chemical Marketing Reporter (Schnell Publishing, weekly) carries reports of expansion plans, mergers, market positions, and government legislation worldwide, with comments from the organizations concerned within the plastics industry.

Chemical Week (McGraw-Hill) publishes the chemical industry leading indicator, predicting trends in USA production of chemicals and allied products.

Chemistry and Industry ('Blue Bits') (semi-monthly, Society of Chemical Industry) contains a short business news section often containing items on plastics industry developments.

The Economist (Economist Publications Ltd., weekly), is an important source of information on current affairs frequently carrying useful reviews in a very readable form.

European Chemical News (ECN) (Business Press International) has a well deserved reputation for high quality reporting on developments in the chemical industry. All sectors likely to be of interest to the plastics manufacturer are reviewed, in the monthly monitors, for example. Downstream processors, with increasing coverage of the global scene are also examined. Full indexes are published half-yearly.

Chemscope is a separate, complementary publication with annual reviews focusing on specific topics, but of wider interest to the chemical industry, for example, Storage and Distribution.

PLASTICS AND RUBBER PERIODICALS

While the business activities of the major plastics producers are a major feature of the coverage, many downstream plastics fabricators' and users' activities are also included. The focus tends to be national, which may limit the interest of some journals, particularly if they are in an unfamiliar language. All tend to have a similar format of regular business features surrounded by advertisements, which may also make scanning difficult. Most journals carry news and comment on companies, production, forecasts, plant investments, new markets, developments and applications, machinery and processes. English summaries are often included.

A selection of titles is given below — there will inevitably, be many omissions — but it is hoped that it will be reasonably representative. Further examples may be found in the 'periodicals abstracted' lists of the major data base producers. Journals containing business information on plastics and rubber include: *British Plastics and Rubber* (monthly, MCM Publishing); *Caoutchoucs et Plastiques* (appearing nine times annually, Soc. d'Expansion Technique et Economique); *Elastomerics* (monthly, Communication Channels); *European Rubber Journal* (monthly, Crain Communications); *Gummi Fasern Kunststoffe* (monthly, Alfons W. Gentner Verlag); *Kunstof Rubber* (monthly, Kunstoffen en Rubber Institut, TNO te Delft); *Materie Plastiche ed Elastomeri* (monthly); *L'Industria della Gomma* (monthly, Italian Association of Rubber, Electric Cables and Associated Industries); *Plaste und Kautschuk* (monthly, VEB Deutscher Verlag fur Grundstoffindustrie, Leipzig); *Plastics and Rubber International* (monthly, Plastics and Rubber Institute) and *Plastics and Rubber Today* (monthly, Maclaren Publishers Ltd.).

Plastics and Rubber Weekly (PRW) (Maclaren Publishers Ltd.,) is highly regarded as an extremely useful and wide-ranging source of business information. *Polymers and Rubber Asia* (bimonthly,

SKC Communications Services, London) is also useful for news of events in the Far East.

A large number of journals are published concentrating exclusively on developments in the plastics industry including titles such as *European Plastics News*, incorporating *British Plastics* and *Europlastics Monthly*, (Reed Business Publishing); *Italian Technology*, (monthly, ERIS SpA); *Japan Plastics Age*, (bimonthly, Plastics Age Co.); *Kunststoffe Plastics*, (monthly, Vogt-Schild AG); *Making it with Plastics* (monthly, MCM Publishing); *Modern Plastics International* (monthly, McGraw-Hill); *Plast rivista delle materie plastiche* (monthly, ERIS, Edizione per l'industria SpA); *Plastics Bulletin* (monthly, a multilingual journal published in Dutch, French and English by Professional Information Media); *Plastics Industry News* (monthly, Institute of Polymer Industry); *Plastics Technology*, (monthly, except semi-monthly in June, Bill Communications); *Plastics World* (monthly, Cahners Publishing); *Plast Panorama Scandinavia* (monthly, Thomson); *Plastverarbeiter* (monthly, Zechner and Huthig Verlag); *Poliplasti e Plastici Rinforzati* (monthly, ETAS Periodici); *Polymer News* (monthly, Gordon and Breach).

More specialised titles include *Adhesives Age* (monthly, Communications Channels); *Advanced Composites Engineering* (quarterly, British Plastics Federation and the Design Council); *Composites* (monthly, Butterworth Scientific); *European Coatings Journal*, the international edition of *Farbe und Lack*, (monthly, Curt R Vincent Verlag); *Imballaggio* (monthly, ETAS Periodici); *PackPlas*, (quarterly, covering the major Arabian countries, International Printing Communications); *Plastics Today: a business magazine for the plastics industry* (ICI); *PSA Plastics South Africa*, incorporating *Plastics and Rubber News* (monthly, George Warman Publications Pty.) and *Swiss Plastics* (monthly, Verlag Dr Felix Wust).

Company information services

Sources of information on companies and their business activities, presented and packaged in different ways to suit the purpose for which they were designed, from checking on financial status to identifying suppliers and potential markets, are rapidly growing in number. Access to basic information, name, address and latest set of accounts for UK registered companies is obtainable from the Companies Registration Office, by direct application or by consulting the microfiche indexes available in the larger business libraries.

Annual reports

Collections of individual company reports containing current financial results and reports of progress with new products and projects are frequently useful sources although it is rare to find a set which is comprehensive (although they are usually freely available), classified, indexed or accessible other than via the company name.

The Science Reference Information Service (SRIS) at Holborn holds one of the largest collections, including those of the top 200 UK companies on open, and a further 2000 on closed access.

Company information from house journals has been covered in the section on Trade Literature.

Other company information services include the *Extel* card service, containing summarized financial and corporate information on UK, European, Middle and Far East regions. *McCarthy* cards containing press cuttings of company activities arranged by company name, and in the case of the UK only, by industry sector. The cuttings, reproduced on 5-column cards in date order, are classified under 13 group headings which are further subdivided to a reasonably specific level, e.g. group F covers chemicals and chemical engineering:

F3/6	Plastics and Resins is further subdivided into
F3/6/1	Polyethylene, polystyrene, polyvinyl chloride, polyamides
F3/6/2	Plastics laminates and composites, polyesters
F4/2	Plastics extruding processes and machinery
F4/4	Plastics mixing machinery

Newspaper sources are important both for current and historical information. One of the most important sources is the *Financial Times* which produces annually cumulated, monthly FT Business Indexes to the wealth of material it contains ranging from specific items concerning a single company to wide ranging surveys of entire industrial sectors.

Other newspapers, e.g. *The Times*, have similar services to offer. Many are available online through data bases such as Textline. Structural changes in the UK polymer-based industries have been reviewed in *Corporate changes in the UK plastics industry — acquisitions, liquidations and new companies* (Corporate Development Consultants, 1982).

Europe's 15000 largest companies 1987 (ELC International) provides ranked lists of the companies by industrial sector, including manufacturers of plastics, resins, rubbers and their products.

Market research

King (1986) draws attention to the existence of two major and distinct types of market research; original, 'field' based research compiled from samples of relevant populations of individuals and firms, and 'desk' research based on published material. Four main categories of publisher are identified, the trade organization, the commercial market research organisation, surveys appearing in newspapers, trade and business periodicals and a miscellaneous group comprising bank, stockbroker, and other groups of a quasi-political nature.

Market surveys are frequently commissioned by individual or small groups of companies, and may be difficult and expensive to acquire, at least not until some considerable time has elapsed after their first appearance. However, they are an information-rich source which cannot be neglected, since they contain collections, analyses and evaluations which may not be obtainable elsewhere.

The largest publicly available collection of market research reports in the UK is held by the Business Information Service of the SRIS. Individual reports are shelved by location mark and a subject index is provided, usefully drawing together information on the range of reports available.

Some of the titles available for reference, taken at random to illustrate the coverage of topics include: *Advanced composites: new engineering plastics* (1982); *Plastics processors* (11th edn, 1985); *Western European plastics and marketing*; *BP — the next 10 years*; *Financial performance in plastics processing*; *Quality of plastics components for TV*; *World plastics to 1995*; *Plastics under the bonnet*.

Bibliographies are published by other organizations, including *Marketing surveys index* (monthly, Marketing Strategies for Industry (UK) Ltd) as a cumulative index to abstracts containing full details of the market survey, including price, supplemented with a complete alphabetical listing of publishers. The index gives alternative subject headings to search, an indication of country coverage, number of pages, report and publisher codes for easy reference to the details in sections printed on white and yellow pages, respectively.

Market research sourcebook (Headland Press) is a comprehensive compilation of source material carefully evaluated and presented, which also offers practical advice on procedures for carrying out market research from scratch.

Marketing surveys index (MSI (UK) Ltd, Mitcham, and the Institute of Marketing, 1987) provide detailed indexes to brief abstracts of the contents of a wide range of reports, studies and surveys.

IAL Consultants Ltd, London produce a wide range of multiclient studies covering individual countries or regions. These are expensive to produce but the shared costs reduce the expense to individual clients. Recent titles include *Advanced engineering materials*: Vol.2, *Thermoplastic composites*, Vol.3, *Thermosetting composites; hot-melt adhesives*. IAL also operate an enquiry service supplying commercial and technical information on all aspects of production, company, financial data, trade statistics, etc.

Jordan and Sons Ltd (London) have recently produced an interesting report on *Britain's plastic industry* (1987), a survey including ranked financial data on 158 companies active in the plastics sector.

Other important surveys include the *ICC business ratios* (a Division of InterCompany Comparisons Ltd., London) which present relationships between sets of financial data in a format allowing proper comparisons between sets of financial performance and growth in companies of different sizes. The full accounts of each company are detailed separately.

Frost and Sullivan Inc., New York, provide market research reports covering many industries, including for example, *Chemicals and Plastics*, a study of market size by product and end-use for the USA and Europe, broken down by country, and *Advanced Composites Materials* covering carbon, boron, and aramide (Kevlar) fibre manufacture for end uses e.g. military and civil aircraft, helicopters, spaceflight, missiles, carbon brake systems and sporting goods.

Another useful compilation covering 11 different sources of information on the UK market for consumer goods and services is the *A-Z of UK marketing information sources* (Euromonitor Publications). Sources discussed range from official government publications, trade associations, private publishers, stockbroker reports, to conventional journals, abstracts and indexes, libraries and information services, computerized data bases and data banks.

Stockbroker reports give the inside story as well as interpreting the raw data. The analysts are in close contact with the policy makers within the companies and often, but not always, the evaluation may be dependent on the share position of the stockbroker in the company. There are many such reports covering individual companies and whole market sectors.

Crawford's directory of city connections is a comprehensive source of information on stockbroking firms and the companies specifying them as principal advisers. Many of the reports are available in the Business Information Section of the SRIS.

European marketing data and statistics (Euromonitor Publications Ltd.) covers 30 countries of Western and Eastern Europe, providing trade and production statistics on all basic marketing parameters, including data on marketing for consumer goods.

An interesting source of export marketing information on inter alia, plastics materials and end-use markets, their share, size and structure is provided by the British Overseas Trade Board Product Data Store, at the Statistics and Marketing Intelligence Library. The information is collected from the foreign trade press, trade associations and bank industrial surveys and collated by embassy staff. Microfilmed reports indexed by country, product and industry code are rapidly accessed by computer for immediate screening and if desired, printing. Material is held for two years before disposal.

Of greater immediacy and potential relevance perhaps is the Export Intelligence Service for UK subscribers seeking market opportunities for products and services abroad. A coded profile of interests is matched daily against the coded market information input received from nearly 200 British Overseas Diplomatic posts, and despatched to the enquirer.

Further help in obtaining overseas market information either from original (possibly grant-aided) research or by purchasing published marketing research may be obtained from the British Overseas Trade Board as part of their export promotion services.

Findex: the directory of market research reports, studies and surveys (FIND SVP, New York), covers 12 major industry categories giving the name of the report, date of publication, number of pages and price and a description of the contents and publisher. Subject and publisher indexes and directory are included.

The National Economic Development Council Committees produce a wide range of reports on the economic performance, problems and opportunities of the industry. The Plastics Processing EDC has produced a series of reports including: *Tooling for shorter injection moulding runs; Plastics in the automotive components markets in Europe; The future of the tyre industry; Manufacturing performance of the process plant industry; Replacing metals with plastics* and *How to succeed in plastics processing*. Plastics Processing EDC reference bulletins and a series of newsletters promulgating the work of the EDC are also produced.

Britain's plastics industry (Jordans and Sons), J. Geach, 1987, covers applications of the major plastics, processes, the structure of the industry, markets, major industrial groups, sources of raw materials and industrial developments.

Statistics

Reading the footnotes to virtually any collection of numerical data, official or otherwise, relating for example to the production, consumption or value of a product, it is not difficult to realise that statistical information is a very complex subject, to be treated by the inexpert with the utmost circumspection, particularly when making comparisons of data from different national sources. Quite apart from the difficulties of consistently classifying the subjects on which the data is to be collected, figures may be subsequently cumulated, estimated, provisional, revised, rounded or simply not available. A brief, but entertaining discussion of some of the traps lying in wait for the unwary is to be found in Huff and Geis, *How to lie with statistics* (Norton, 1954).

The British Plastics Federation has pointed out that statistics for the plastics processing industry are notoriously difficult to compile (*International status report on plastics 1982*), principally owing to in-house production by user industries and the difficulty of estimating the plastic content of articles which are not classified as plastics products.

There is however, no shortage of statistical sources, many arising from government activities and published at regular intervals. The *Central Statistical Office's guide to official statistics* published biennially by HMSO contains 16 chapters, 120 sections and over 800 sub-sections on sources of government and non-government statistics for the UK. Under Chemicals and Allied Industries, Section 7.9 (i) lists four sources:

a) *Business Monitor PQ276.1, Synthetic Resins and Plastics Materials*;

b) the *Monthly Digest of Statistics*, a quarterly analysis of production of products of condensation, polycondensation, poly-addition, (co)polymerisation, cellulosics and other plastics and modified natural resins, intermediate forms between resin and semi-fabricated stages and semi-finished products;

c) a reference to *British Business* quarterly series *Activity in the Chemicals Industry* giving data on output, sales, exports, imports, home market price indices of plastics and synthetic resins and

d) the *Annual Abstract of Statistics for UK production* of synthetic resins by type, exports and imports and an area breakdown.

Similar series are listed for Synthetic Rubber including *Business Monitor PQ 491, Rubber*, the *Monthly Digest of Statistics*, items in *British Business* (weekly, Dept. of Trade and Industry) and the

Rubber Statistical Bulletin, a monthly source of data on production, exports, imports, consumption and stocks. The index yields further sources under the subject headings of Plastics, -inflatables, -materials, -products and -protective footwear.

The *Business Monitors* are a series of statistical publications produced quarterly by the Business Statistics Office with the co-operation of firms which provide unique data on production, sales, employment and investment.

About 300 *Production Monitors* are produced quarterly including *PQ 4831 Plastics Coated Textile Fabrics; PQ 4832 Plastics Semi-manufacturing; PQ 4833 Plastics Coated Floor Coverings; PQ 4834 Plastics Building Products* and *PQ 4836 Miscellaneous Plastics Products*. A similar series of Annual Census of Production Reports is also produced e.g. *PA 481 Rubber Products* and *PA 483 Processing of Plastics*.

Business Monitors for rubber include *PQ 4811 Rubber Tyres and Inner Tubes; PQ 4812 Miscellaneous Rubber Products; PQ 4820 Retreading and Specialist Repairing of Tyres*.

Help in finding further information on statistics may be obtained from the Information Services Division, Great George Street, London, from the Business Statistics Office, Cardiff Road, Newport, Gwent or from the appropriate Government Department concerned (each Government Department is responsible for publishing its own directories).

Digests of UK Energy Statistics include energy consumption figures by the plastics and synthetic rubber industry.

Many of the UK serials have their national equivalents in other countries throughout the world, including the *Council for Mutual Economic Assistance (CMEA) Statistical Yearbook* containing data on the USSR, Bulgaria, Hungary, East Germany (DDR) and other COMECON countries, the *Gesamtverband der Textilindustrie*, West Germany (FDR), and the Institute Central di Statistica *Annual del Comercio con l'Estero* in two volumes, with import and export data for commodities, classified by the BTN classification. *Annuario di Statistiche Industriali* is the annual compilation. Other national series include the Netherlands *Central Bureau voor de Statistiet Produktstatistiken* for the motor, chemical, synthetic materials processing and rubber industries. The US Department of Commerce: Bureau of the Census produces a *Census of Manufacturers: Industry Series* covering plastics materials and resins, synthetic rubber, cellulosics and man-made fibres, adhesives and sealants, tyres and inner tubes and rubber and plastics footwear. Other series include *Imports for Consumption* and *General Imports* (includes polyesters and phenolic resins) and *Synthetic Organic Chemicals: USA Production and Sales*, the major US government publication on plastics. Japanese official sources include the *Ministry of Trade and Industry (MITI) Yearbook of Rubber Statistics*.

Supranational organizations such as the European Economic Community (EEC) and the United Nations (UN) are major producers of statistical material, the *UN Statistical Office Monthly Bulletin of Statistics* for example, being a source (in French and English) of individual countries' monthly production figures for the carefully defined Plastics and Resins sector. These update the *UN Yearbook of International Trade Statistics. Commodity Trade Statistics* gives annual figures (converted to US dollars and metric units for strict compatibility) on world community trade classified by the Standard International Trade Classification (SITC), allowing comparisons to be made with reasonable confidence. SITC is fairly specific for the major plastics types, polypropylene and polyvinyl chloride, but inevitably for the sake of consistency some products have to be lumped together, and the detail becomes blurred. A further UN bulletin, the *Annual Bulletin of Trade in Chemical Products* produced by the Economic Commission for Europe, Geneva gives worldwide figures for origin and destination of addition and condensation polymer products.

Worldwide rubber statistics published by the International Institute of Synthetic Rubber Producers gives data on production broken down by rubber type, consumption and on the motor trade, vehicle and tyre production. A further useful source is the *Rubber Statistical Bulletin* produced by the International Study Group, London.

Chemical and Polymer Production Statistics (monthly), the *World Plastic Service* and the *Chemical Forecasting Service* (quarterly) are comprehensive publications available from Data Resources Inc., McGraw-Hill, Stockport, Cheshire.

References

Huff, D. and Geis I. (1954), *How to lie with statistics* (Norton).

King, D.M., (1984), Market Research Reports in *Guide to sources of information in management and business*, pp 189–200 (Butterworth Scientific Ltd).

Smith, J.K., (1985), 'How to write business intelligence reports', *Business Information Review*, **1** (4), April, 11–22.

Tattum L. (1987), *European Chemical News*, **48** (1258), January 19, 17–18.

CHAPTER THIRTEEN

Fibres

R. J. E. CUMBERBIRCH

Introduction

This chapter sets out to indicate how and where main sources of information on fibrous materials may be found. Fibres are the raw material of the textile industry, which converts them into yarns and fabrics and applies mechanical or chemical treatments to modify and enhance their properties so that they may perform more effectively in particular applications. The 'finished' yarns and fabrics produced by the industry are subsequently supplied for conversion to apparel, domestic textile, furnishing, and industrial, medical and other specialised fibre-containing products.

A large volume of information on fibre materials and their properties, processing, and uses stems, of course, from the textile industry. But a large additional volume of relevant information comes from areas of activity allied to the textile industry, for example, from the agricultural and chemical industries. Natural fibres such as cotton, wool, silk and flax are grown and harvested by agriculture and much of their quality and properties depend on conditions and geographical regions of growth and means of harvesting. Man-made fibres are produced by the chemical industry, either via the chemical treatment of natural polymers such as cellulose or the synthesis of fibre-forming polymers from simple monomers. The chemical industry also supplies the wide range of agents and products used to treat or modify fibre materials; bleaches, dyes, flame retardants, resin finishes and polymer coatings are a few examples.

Other areas of industrial activity that give rise to information of major importance to fibre science and technology are those concerned with the development and manufacture of particular types of fibre- or textile-containing products and, in particular, with the assessment of the performance of these products. The best known and most obvious example is the clothing industry which produces a wide range of garments ranging from underwear, etc., through outerwear to specialist protective clothing for industrial and other workers. But there are many other areas which are less known although of very great importance to textile science and technology. These include areas concerned with industrial textiles (e.g. tarpaulins, tentage, papermakers' felts, protective covers), fibre-reinforced composites (e.g. tyres, conveyor and driving belts, cement reinforcement, high-strength structural components), ropes and cordage (cables, hawsers, twines), medical textiles (e.g. suturing, surgical gowns, prostheses, sanitary wear), filtration (use of fibrous materials in filters), geotextiles (use of fibrous materials to reinforce and drain soil-based or other civil engineering structures), agricultural and horticultural materials (e.g. screening), and military and other specialised equipage. In all of these areas the high- or special-performance characteristics of fibre materials are of paramount importance (as a mountaineer, a wearer of bullet-proof clothing or a parachutist would no doubt confirm!). Because of the focus of interest in today's science and technology on 'high-tech' and high-performance materials, information on fibre properties and performance that stems from such sectors is of particular importance and is frequently sought.

The upshot of this preamble is that sources of information on fibre materials, their properties, performance and uses are widespread because the use of fibre materials is widespread. This means that a searcher for information has to be very much aware of this fact and has to be prepared to search 'widely' and often outside the purely textile field for some types of information.

This review will assume that 'fibre information' encompasses the broad range of textile information since the polymers and plastics industry and the users of this industry's products make considerable use of or apply polymeric materials to a wide range of textiles in the manufacture of end products.

II Sources of information on fibrous materials

(a) Books

As in any technology, textile technology gives rise to a constant and considerable flow of new books or new editions of existing books, all of which add to a large quantity of existing books. Such publications range from craft and history books, through students' books and general reference books, to books on the detail of textile and related science. No single comprehensive listing of all such books exists but there are two useful sources of information on many of these publications.

The Textile Institute (1) publishes a large number of books on textile materials and textile technology and also acts as an agent in the sale of similar books which have been published elsewhere. It produces listings of the wide range of books it offers for sale and these listings are available for consultation.

The British Textile Technology Group (formerly Shirley Institute) (2) records book publications in *World Textile Abstracts* (see under 'book' in the annual subject indexes) and the associated *World Textile Abstracts data base*. The latter can be accessed online via the DIALOG (3) computer-based information systems. The citation of the book-identifying keyterms ELEMENTARY BOOKS, REFERENCE BOOKS, CONFERENCE PAPER BOOKS, PATENT REVIEW BOOKS and DIRECTORIES (BOOKS), as appropriate, plus, if desirable, other keyterms specifying the relevant type of subject matter, will list bibliographic references to some or all of the several hundred complete books recorded since 1970. Alternatively, use of the 'type of literature' field in the data base and citation of the appropriate code (4) for 'books', plus citation of appropriate keyterms to identify relevant subject matter, will produce a listing of complete books *plus*, in addition, chapters in books (e.g. books on 'polyester fibres' and books which contain a chapter on polyester fibres).

A few books are worthy of brief mention in this chapter because of their value as general reference sources or because of their likely usefulness to information seekers from the polymers and plastics field.

Textile terms and definitions (5) is a book issued in up-dated editions which lists the terminology of the textile industry and also provides internationally-accepted definitions of the individual terms, often with explanatory notes and diagrams.

Identification of textile materials (6) provides detailed information about the character, properties and appearance of textile fibres and the means of identification of individual types.

Fiber chemistry (7) gives detailed accounts of the composition, structure, properties and chemistry of the main natural, regenerated and synthetic-polymer fibres and *Man-made fibres* (8) is a useful reference work on the various types of regenerated- and synthetic-polymer fibres, their properties and uses.

Strong fibres (9) gives detailed information about the production, structure, properties and uses of inorganic, carbon and high-tenacity organic fibres which find main application in present-day fibre-reinforced-composite technology.

A 'fibres' book also worthy of special mention is *Carbon & high performance fibres directory* (10) which lists the manufacturers, types and properties of carbon, aramid, glass, silicon carbide and other fibres, fabrics and prepregs as designed for use in fibre-reinforced composite materials.

Last in this brief selection is *The standard handbook of textiles* (11), probably the best and most authoritative general reference book on the textile industry and its range of manufacturing processes.

(b) Directories

The demand for knowledge of 'who produces what' is on the increase but sources of comprehensive information in this area tend to diminish because of the increasing cost of and expertise required in collecting and recording such information and the increasing rate of change in industry brought about by closures, mergers, changes of product line and the like. Such changes have been a particular feature of the textile industry in recent years.

Each national industry gives rise to directories of various types and degrees of coverage, and it is usually not at all easy for an information seeker to establish a reasonably comprehensive and up-to-date list of the suppliers of a particular type of textile material, instrument or machine in a given country or region.

The USA textile industry is well covered by *Davidson's salesman's book* (12). The UK textile industry is best covered by the *Kompass buyers' guide, textile & clothing* (13). A world listing of manufacturers of man-made fibres is published annually in the June issues of *Textile Organon* (14), a monthly periodical providing statistical data on the USA and world textile industries.

Textile-relevant directories can be listed by use of the *World Textile Abstracts* data base in DIALOG and the citation of the keyterms DIRECTORIES (BOOKS) or DIRECTORY INFORMATION. The latter keyterm covers compiled collections of information on manufacturers and products such as are published in special periodical issues (e.g. buyers' guides) or as part of the content of a periodical article. Both these complementary keyterms would be cited in a particular search (e.g. DIRECTOR?/DE) together with other keyterms specificying the broad subject area, e.g. MANUFACTURERS and (MAN MADE or SYNTHETIC) for manufacturers of man-made and synthetic fibres. It is often worthwhile for information service and general references purposes to obtain a printout list of all directory information as published (say) during the last 3 years. This listing can then be readily consulted when the need arises for a particular type of directory information.

(c) General literature

The broad mass of information on which science and technology feeds and develops is made available to the world at large through a variety of publications: articles in scientific, technical and 'trade' periodicals, patents, technical brochures and reports, books, monographs, theses, standards, etc. Abstracting services diligently search out the presence of information in such publications and publish periodicals and annual indexes which review and record these sources of published information. Each of these services covers literature published in a broad subject-defined field, and there is frequently considerable overlap between fields; this can result in a particular item of published information being recorded by two or more abstracting services.

In today's computer-based world, such abstracting services also usually enter the records they produce on magnetic tape or disk to form data bases which are then entered in one or more 'centralised' computer-based information systems for public access and use. These systems enable anyone, in most parts of the world, to search 'online' any one or more of the data bases mounted within each system and to identify sources of information that are likely to be relevant to his or her need. The information published in the sources so identified can subsequently be obtained in the form of library-held literature or as photocopied items obtained by the use of appropriate facilities.

Let us now see how abstracting systems serve the 'textile' information seeker.

ABSTRACTS PERIODICALS

There are two well established abstracts periodicals which serve the world's textile information needs: *World Textile Abstracts* (15) published by the British Textile Technology Group in the UK and *Textile Technology Digest* (16) published by the Institute of Textile Technology in the USA. The *World Textile Abstracts* coverage tends to be concentrated on the more 'solidly' scientific and technical literature of relevance to fibres and textiles, and provides coverage of European, UK and USA patent literature. *Textile Technology Digest* tends to cover more thoroughly the trade and news literature and the detail of clothing and related manufacture. But both periodicals provide extensive coverage of the broad mass of textile literature, the coverage of each very considerably overlapping that of the other. The annual subject indexes published in *World Textile Abstracts* are particularly useful when there is a need to trace literature on a particular subject which was published during a period of, say, one or two years.

It should be noted that a list of periodicals published throughout the world which have at least some textile relevance is presented in the first issue of *World Textile Abstracts* each year.

In addition to these two textile-literature covering periodicals, there are several other abstracts periodicals which cover published information of relevance to fibres and textiles. These include: *Chemical Abstracts (Macromolecular Section)* (17) and *RAPRA Abstracts* (18) which cover the synthesis, properties and uses of fibre-forming polymers and polymers with application to textiles, and *Abstract Bulletin of the Institute of Paper Chemistry* (19), *Paper and Board Abstracts* (20), *Printing Abstracts* (21) and *International Packaging Abstracts* (22) which cover paper and related science and technology, many facets of which are close to or overlap textile science and technology.

DATA BASES ONLINE

The publishers of all the abstracts periodicals noted also produce equivalent data bases for use within computer-based information services. Thus, *World Textile Abstracts* can be searched by means of the DIALOG system (3), in which it is listed as 'file 67'. *Textile Technology Digest* can be searched by means of the DIALOG system in which it is listed as 'file 119'.

The basic features of these data bases are similar to those of most of the other data bases held within the computer-based information systems. Every data base has a number of record fields in each of which is contained a particular type of record: titles of published items, author names, source publication, language of publication, abstract, special indexing and so on. Individual record fields can be selected for use in searches according to need. For example, a search might require the listing of references which meet the specification: items of literature published by one or more cited authors (using the author record field), in a cited language (using the language record field), and indexed by cited subject-identifying terms (using the record field which contains terms indexing subject matter). The structure of record fields and their use in online information searching is covered in some detail elsewhere in this book and need not be explained again at this point. It should be noted, however, that each data base has its own characteristics and features and, if it is to be used fairly frequently in online searching, these should be studied and made use of in order to obtain the best search results.

The *World Textile Abstracts data base* has a number of special features which can be of particular value in certain types of information search. Those relating to the production of book listings and listings of sources of directory information have already been mentioned.

Another type of listing which is not infrequently required is that of sources of statistical data. Such data is usually published in two basic forms: as regularly issued compilations of 'official' statistics (e.g. *Industrial Production* published quarterly by Eurostat) and as compilations of statistics presented in periodical articles by their authors. The former can be listed by citation of the keyterm STATISTICS REPORTS, in association with keyterms specifying a country (e.g. USA) or region (e.g. WESTERN EUROPE) if so required. Anyone having a need for textile statistical data from time to time can be advised to produce such a listing via the *World Textile Abstracts data base* so as to have a readily available list of references to these official sources. The latter (irregularly published sources of statistics) plus the former (regularly published sources) can both be listed by the citation of the keyterm DATA. This term indexes numerical data that relate to a particular country or region, e.g. Japanese consumption, production and trade data. A few sources of such data are of historical value only and, if thought desirable, may be eliminated from any listing by the citation of 'not HISTORY' in a search command. An example of search terms might be: DATA and USA and (MAN MADE or SYNTHETIC) and (PRODUCTION RATE or INTERNATIONAL TRADE).

The *World Textile Abstracts data base* covers all relevant British, International and United States standards and these can be listed by citation of the keyterm STANDARDS plus the country (e.g. USA) or the 'WORLD' keyterm according to need. As in any other type of search, broad subject matter (e.g. FLAME RESISTANCE) may also need to be specified, e.g. STANDARDS and USA and FLAME RESISTANCE.

The patents issued by the British, European and United States Patent Offices are also covered by the data base. In recent years the International Patent Classification assigned to each patent has been recorded at the end of each abstract. Consequently the relevant codes (e.g. DO2G) are treated as subject-indexing 'words' and form part of the basic subject index which is made up of words extracted from titles and abstracts plus keyterms. Patents classified by a given code can be listed in a search operation by citation of the code, e.g. S DO2G.

General subject matter is indexed in the *World Textile Abstracts data base*, as in other data bases, in two ways (i.e. in two record fields): by the individual words which have occurred in titles, abstracts, and assigned descriptors or keyterms (i.e. as are recorded all together in the basic subject index field noted above), or by keyterms, formed from one or more words, which have been specially assigned for indexing purposes (i.e. as are recorded as full terms in the 'descriptor' field). The latter are listed in a *Register of keyterms* (23). Because of the confusion in or casual use of terminology which can occur, particularly in the field of technology, it is strongly recommended that online searches conducted on a subject basis make use of the keyterm indexing or 'descriptor' record field rather than the basic index field. This requires reference to the *Register of keyterms* for identification of appropriate keyterms to be cited in a search. It should be noted that, in computer preparation of the data base, the computer ensures that only the 'preferred term' from a possible choice of equivalent or near equivalent terms is entered. It should also be noted that, at the time of data base preparation, the computer may add additional, 'broader', keyterms to those already listed according to the instructions given in the *Register of keyterms*.

In general, it is recommended that online searching of the *World Textile Abstracts* or other data bases for sources of subject-specified information should be conducted on a reasonably broad basis. This will considerably increase the likelihood that all relevant sources will be listed and reduce the chance of omissions. It is often believed, consciously or unconsciously, that a computer-based information retrieval system is able to identify with accuracy and absolute certainty the exact subject-identified references which are required by a searcher. For a number of reasons, this is not so. First, the whole system depends on the words or codes which have been recorded to index each item of literature and it is neither practicable or possible to ensure each and every item has been appropriately and perfectly indexed at the time of recording. Secondly, the computer cannot prejudge the value to the searcher of each item of literature it selects for listing: one searcher may judge an item to be irrelevant and valueless whilst another, with the same general information requirement, may judge it to be relevant and valuable. Thirdly, each subject-indexing word or keyterm recorded to index an item of literature cannot also measure the amount of information in the item so indexed: the amount may be small or it may be large and represent the whole focus of attention of the item concerned. Fourthly, information seekers themselves are often uncertain about the scope of the information they seek and, as with browsing along library shelves in a search for information, the act of conducting a search online often results in the information seeker altering and/or extending the specification of his or her information requirement.

Thus, it is considered desirable that searches be conducted on as broad a subject base as is reasonable and appropriate: it does not take long to scan and eliminate 'unwanted' references from a computer printout but it can be highly disadvantageous (and sometimes embarrassing) if a restricted and over-precise subject-based search specification has blocked the listing of references which would have, in fact, been highly relevant to an information requirement.

The extent and amount of a 'broad-based' listing produced by online subject-based retrieval can, however, often be limited by the appropriate use of other record fields. Thus, for example, are patents to be excluded or should the search be restricted just to patents (use of the 'document type' field)? Should literature in all languages be listed or just those items in English, or in English and German, etc. (use of the 'language' field)? Should the search be limited to sources of information recorded in a limited time span, e.g. the last 5 years (use of the 'recording year' field)? Such search fields form part of most data bases in the DIALOG and INFOLINE information services, but the individual fields are identified according to the particular codes adopted by the data base producers and/or the information services which mount the data base for world-wide access. Consequently data base-producers' and information-service documentation has to be consulted in order to identify the correct codings and form of search specification for use in online searches.

(d) Standards

Reference has already been made to the use of the *World Textile Abstracts data base* for the listing of standards, particularly British, international and United States standards. Since there are a very large number of standards which relate to the testing and performance of textile and textile-containing materials, and since official listings of standards are not always helpful in enabling information searchers to track down standards relevant to a particular need, an online search of an appropriate data base is often a useful and speedy means of meeting a need to identify particular types of test or tests on a particular type of material.

The British Standards Institution used to issue quite a well-known handbook (*Methods of test for textiles*) which contained a collection of the British Standards that were basic for the textile industry. However, up-dated editions of this handbook are no longer published and so the last (1974) edition has become increasingly out of date.

The situation in respect of United States standards is much better. The American Society for Testing and Materials (ASTM) publishes annually up-dated volumes (24) which contain new, revised and existing standard test methods promulgated by this society and which are of relevance to particular areas of technology. Two of these volumes relate to textile materials. The American Association of Textile Chemists and Colorists (AATCC) also publishes annually a compilation (25) of its new, revised and existing standards, all of which relate to chemical, colour and related forms of textile testing.

General enquiries about standards should be made to the appropriate national or international offices (e.g. 26, 27, 28).

(e) Expert organisations

Technology and new products are developed at an increasing pace in today's world and much of the information generated either takes considerable time to reach the printed page (and hence to become accessible to the world in general) or is kept confidential to the generators in order to preserve commercial advantage as long as possible. Many of the requests for information made to information services are, quite naturally, concerned with the latest technology, but these requests often cannot be satisfied by citation of printed sources except at the most general level.

Unless private contacts in the field of the relevant technology are available to the information seeker, the best hope of obtaining at least some information is by making requests to appropriate expert organisations. Often these organisations will provide 'off the cuff' general information or, at least, will redirect the enquiry to a more appropriate body. They will usually also provide more detailed information (if they have the necessary expertise) but are likely to request a fee to compensate them for the time and effort involved in its provision.

Therefore, to complete this chapter on textile information sources, it is worthwhile noting UK organizations that may be able to provide specialist information on an advisory or technical basis (it should be noted that a similar range of organisations available for technical consultation exists in many other countries).

RESEARCH ASSOCIATIONS

British Textile Technology Group (2): textile science, technology and product applications
Fabric Care (29): laundering, dry cleaning, fabric aftercare
Hatra (30): hosiery and knitting
Lambeg Industrial (31); flax, polyolefin technology
RAPRA Technology (32): rubber and plastics

UNIVERSITIES (WITH TEXTILE EXPERTISE)

Bradford (33): wool and related textile technology
Leeds (34): man-made fibre and wool technology
Loughborough (35): engineering aspects of textile technology
Strathclyde (36): fibre properties and geotextiles
UMIST (37): general textile technology

COLLEGES (WITH TEXTILE EXPERTISE)

Bolton Institute of Technology (38): cotton and related technology
Huddersfield Polytechnic (39): textile marketing
Leicester Polytechnic (40): knitting technology
Scottish College of Textiles (41): textile colour and computing technology
Trent Polytechnic (42): knitting technology

PROFESSIONAL BODIES

British Man-made Fibres Federation (43): man-made fibre manufacture
British Textile Employers Association (44): sources of textile materials
Society of Dyers and Colourists (45): colour technology
Textile Institute (1): general textile technology

References

(1) Textile Institute, 10 Blackfriars Street, Manchester M3 5DR, UK; telephone (061) 834 8457, Telex 668297.
(2) British Textile Technology Group, Shirley Towers, Manchester M20 8RX, UK; telephone (061) 445 8141, Telex 668417.
(3) DIALOG Information Services, Inc.: (USA) Customer Services Dept., 3460 Hillview Avenue, Palo Alto, California 94304; telephone (415) 858 3810, Telex 334499: (Europe) Learned Information Ltd/DIALOG, Woodside Hinksey Hill,

Oxford OX1 5AU, UK; telephone (0865) 730 275, Telex 837704.
(4) See *Online to textile literature* published by and available free of charge from British Textile Technology Group — see reference (2)
(5) Beech, S.R. Farnfield, C.A. Whorton, P. and Wilkins, J.A. (eds) (1986) *Textile terms and definitions* (Textile Institute, Manchester) (ISBN 0 900739 82 7).
(6) Farnfield, C.A. and Perry, D.R. (eds) (1975) *Identification of textile materials* (Textile Institute, Manchester) (ISBN 0 900739 18 5).
(7) Lewin, M. and Pearce, E.M. (eds), *Fibre chemistry* in *Handbook of fiber science and technology*, (1985) (Marcel Dekker, New York) (Vol IV) (ISBN 0 8247 7335 7).
(8) Moncrieff, R.W. *Man-made fibres* (1975) (Butterworth, London) (ISBN 0 408 00129 1).
(9) Watt, W. and Perov, B.V. (eds) *Strong fibres* in *Handbook of composites*, (1985) (North-Holland, Amsterdam) (Vol 1) (ISBN 0 444 87505 0).
(10) Pamington, D. *Carbon & high performance fibres directory* (Pammac Directories, London) (ISSN 0268 0491).
(11) Hall, A.J. *The standard handbook of textiles* (1975) (Newnes-Butterworth, London) (ISBN 0 408 70458 6).
(12) Davidson's salesman's book (Davidson Publishing Company, PO Box 477, Ridgeway, New Jersey 07451, USA).
(13) *Kompass buyer's guide, textiles & clothing* (Kompass Publishers Ltd,. Windsor Court, East Grinstead House, East Grinstead, West Sussex RH19 1XD, UK) (ISBN 0 86268 096 4).
(14) *Textile organon* (The Textile Economics Bureau Inc., 101 Eisenhower Parkway, Roseland, New Jersey 07068, USA) (e.g. see 1986, **57**, No.6, 116–160) (ISSN 0040 5132).
(15) *World Textile Abstracts* (British Textile Technology Group, Manchester): 24 issues per annum plus annual index issue (ISSN 0043 9118).
(16) *Textile Technology Digest* (Institute of Textile Technology, PO Box 391, Charlottesville, Virginia 22902, USA) 12 issues per annum (ISSN 0440 5191).
(17) *Chemical Abstracts Macromolecular Section* (American Chemical Society, 2540 Olentangy River Road, Columbus, Ohio 13202, USA) (ISSN 0009 2274).
(18) *RAPRA Abstracts* (Pergamon Press plc, Headington Hill Hall, Oxford OX3 0BW UK) (ISSN 0033 6750).
(19) *Abstract Bulletin of the Institute of Paper Chemistry* (Institute of Paper Chemistry, 1043 E.South River Street, PO Box 1039, Appleton, Wisconsin 54912, USA) (ISSN 0020 3033).
(20) *Paper & Board Abstracts* (Pergamon Journals Ltd., Headington Hill Hall, Oxford OX3 0BN, UK) (ISSN 0307 0778).
(21) *Printing Abstracts* (Pergamon Journals Ltd. Oxford) (ISSN 0031 109X).
(22) *International Packaging Abstracts* (Pergamon Journals Ltd., Oxford) (ISSN 0260 7409).
(23) *Register of keyterms* (British Textile Technology Group, Manchester, UK).
(24) *Annual book of ASTM standards* (Volume 07.01 Textiles — yarns, fabrics and general test methods (e.g. 1988: ISBN 0–8031–1123–1); Volume 07.02 Textiles fibres, zippers (e.g. 1988: ISBN 0–8031–1124–X); (American Society for Testing and Materials, Philadelphia, Pennsylvania 19103, USA).
(25) *Technical Manual of the American Association of Textile Chemists and Colorists* (e.g. 1988: Volume 63) (American Association of Textile Chemists and Colorists, PO Box 12215, Research Triangle Park, North Carolina 27709, USA).
(26) American National Standards Institute, Inc., 1430 Broadway, New York, NY 10018, USA.
(27) British Standards Institution, Linford Wood, Milton Keynes MK14 6LE, UK.
(28) International Organisation for Standardisation, 130 Central Secretariat, 1 rue de Varembe, 1211 Geneve 20, Switzerland.

(29) Fabric Care Research Assocation, Forest House Laboratories, Knaresborough Road, Harrogate, North Yorkshire HG2 7LZ, UK.
(30) Hatra, 7 Gregory Boulevard, Nottingham NG7 6LD, UK.
(31) Lambeg Industrial Research Association, The Research Institute, Lambeg, Lisburn, Co. Antrim BT27 4RJ, Northern Ireland.
(32) RAPRA Technology Ltd, Shawbury, Shrewsbury, Shropshire SY4 4NR, UK.
(33) Bradford University, Bradford, West Yorkshire BD7 1DP, UK.
(34) Leeds University, Leeds LS2 9JT, UK.
(35) Loughborough University of Technology, Loughborough, Leicester LE11 3TU, UK.
(36) Strathclyde University, Glasgow G4 0NS, UK.
(37) University of Manchester of Science and Technology, Sackville Street, Manchester M60 1QD, UK.
(38) Bolton Institute of Technology, Deane Campus, Deane Road, Bolton BL3 5AB, UK.
(39) Huddersfield Polytechnic, Queensgate, Huddersfield HD1 3DH, UK.
(40) Leicester Polytechnic, PO Box 143, Leicester LE2 3TE, UK.
(41) Scottish College of Textiles, Netherdale, Galashiels TD1 3HF, UK.
(42) Trent Polytechnic, Burton Street, Nottingham NG1 4BU, UK.
(43) British Man-made Fibres Federation, 24 Buckingham Gate, London SW1E 6LB, UK.
(44) British Textile Employers Association, Reedham House, 31 King Street West, Manchester M3 2PF, UK.
(45) Society of Dyers and Colourists, PO Box 244, Perkin House, 82 Grattan Road, Bradford BD1 2JB, UK.

CHAPTER FOURTEEN

Rubber

K. P. JONES

Rubbers are highly distinctive polymers due to their characteristic behaviour and to the natural origin of the paradigm material. Rubbers are more complex to process than thermoplastics, and need to be softened to enable other ingredients to be incorporated and permit forming; most require vulcanization. The tyre is the major outlet for general purpose rubbers. There are also self-curing, powdered, liquid, thermoplastic and latex rubbers which do not fully conform to this pattern.

The 'rubber literature' is extensive in time, moderately well documented, relatively sparse, and diverse, extending from the botany of rubber-bearing plants to the synthesis of special-purpose elastomers. To an increasing extent the literature forms an integral part of the polymer literature. Thus *Journal of Polymer Science, Journal of Applied Polymer Science, Macromolecules*, the *International Journal of Polymer Science and Technology, Plastics and Rubber International* (for full details see Chapter 3), and the major monographs on polymers form an essential component within the rubber literature. This is due to the nature of synthetic rubbers, the introduction of thermoplastic rubbers, and the merging of professional bodies and educational facilities: a process which is less evident in the USA. To save space, abbreviations are used for several organizations: the following table provides a key.

Abbreviation	*Full form*	*Country*
ACS	American Chemical Society	USA
ASSOGOMMA	Associazione Nazionale fra le Industrie della Gomma	Italy
ASTM	American Society for Testing and Materials	USA
BSI	British Standards Institution	UK
DKG	Deutschen Kautschuk-Gesellschaft	W. Germany
IISRP	International Institute of Synthetic Rubber Producers	US base
IRRDB	International Rubber Research and Development Board	UK base
IRSG	International Rubber Study Group	UK base
ISO	International Organisation for Standardisation	
MRPRA	Malaysian Rubber Producers' Research Association	UK
MRRDB	Malaysian Rubber Research and Development Board	Malaysia
PRI	Plastics and Rubber Institute	UK
RAPRA	RAPRA Technology (formerly: Rubber and Plastics Research Association)	UK
RRIM	Rubber Research Institute of Malaysia	Malaysia

There is a lack of a recent general introduction: Le Bras (1968) is now dated and Sowry (1977), intended for children, pays slight attention to synthetic materials. There are several recent handbooks: Barlow (1988) and Brydson (1988) both gain in coherence from single authorship. Brydson is more comprehensive and better indexed, but neither book contains sufficient references, although Brydson is once again better. Both Morton (1987) and Blow and Hepburn (1983) are multi-authored and have appeared in several editions. The latter is well structured and includes information on the manufacture of major products: tyres, belting, hose and footwear, and on testing and specifications: there are many references and a 22 page three-column index. Morton (1987) is virtually a new book. It is strongest on synthetic rubbers, but sometimes suffers from bias towards specific products (the chapter on thermoplastic rubbers tends to accentuate block copolymers, for instance). It also includes latices.

Morton (1987) is less thoroughly structured: it includes latex, however. Eirich (1978), published in association with the ACS Rubber Division, is intended to complement Morton (1987). It incorporates contributions from well known authorities on elasticity, polymerisation, rheological behaviour, reinforcement and strength. Porter (1979) noted the absence of oxidation and considered the material on blends, thermoplastic rubbers and chemical modification to be out-of-date.

These may be supplemented by reference to Vanderbilt (1978) which has long enjoyed a reputation for practical usefulness. Freakley (1985) amplifies some of this and also extends to process control. There are many aspects of rubber technology which are still only covered in-depth by Genin (1956–1962), a multi-volume encyclopaedia. Evans (1981) is selective in coverage and may be seen as a precursor of the *Developments in rubber composites* (irregular, Applied Science) series. Although the Small Business (1983) publication has a slant towards Indian requirements, it does incorporate much neglected elsewhere.

Roff and Scott (1971) is the best of the polymer data books for coverage of rubbers: the data are presented on the basis of material rather than property. In addition to physical, chemical and electrical properties, there are concise details on fabrication, uses and history.

The physics of rubber was examined by Treloar (1975), first published in 1949. In an obituary notice, Berry (1987) called it 'a classic in its field'. Elasticity is discussed by Mark and Lal (1982). The physics and chemistry, mainly of natural rubber, is surveyed in Bateman (1963), a multi-author review published to mark the Silver Jubilee of MRPRA: a companion by Roberts (1988) was prepared for the Golden Jubilee. Brydson (1978) has produced a major survey of rubber chemistry which is partly based on chemical processes, such as oxidation, and partly on polymer type. Both Smith (1979) and Barnard (1979) were critical of the lack of adequate referencing. Naunton (1961) contains several major contributions, but suffers from an inadequate index.

Considering that vulcanization forms an almost inescapable element in most rubber processing, it is surprising that so few monographs exist. Hofmann (1967) published an extensive (over 1100 references) review, but this needs to be up-dated by later reviews in the periodical literature, e.g. Kirkham (1978). Also on vulcanisation there is Alliger and Sjothun (1964) and Hills (1971) specifically on heat transfer. Monographs on other specific techniques are also rare, but Brown (1979, 1986) on physical testing, Wake (1983) on chemical analysis, Wheelans (1974) and Penn (1969) on injection moulding, and Wake (1982) on textile reinforcement should be noted.

History

Rubber lacks a comprehensive historical study, but Collins (1987) offers a coherent picture of the commercial development of the industry. Schidrowitz and Dawson (1952) is multi-authored: some contributions are excellent. There is a chronology, but the index is incomplete and there are major gaps. Dean (1987) gives a scholarly account of the rise of the industry in Brazil. Both Blow and Hepburn (1982) and Roberts (1988) contain brief historical chapters.

Drabble (1973) described the early history of rubber in Malaysia; Barlow (1978) is also, in part, a historical study. Woodruff (1958) gave a scholarly account of the British rubber industry in the last century. Seymour (1982) contains excellent brief histories of synthetic rubber by Morton and natural rubber by Hurley. Whitby (1952) contains an extensive history of the development of synthetic rubber by Dunbrook, which may be amplified by Phillips (1963) and Herbert and Bisio (1985) who examine the rapid rise of synthetic rubber output during the Second World War, and McMillan (1979) who recalls the introduction of stereoregular polymerisation by Ziegler and Natta. Some of the early contributors left published records of their work: Hancock (1857) made many advances in rubber technology; Goodyear (1855) discovered vulcanization, and Wickham (1908) collected rubber seeds in Brazil for transfer to Asia.

Literature guides

There have been several earlier guides, notably those by Yescombe (1968, 1976), which record the monograph literature and periodical titles, but fail to assess the documents listed. Yescombe (1976) does, however, attempt to list some periodical articles. Unfortunately, the guides are poorly indexed and the absence of an author index makes reference difficult. There is also an extensive UNIDO (1979) annotated bibliography, which lists organizations, sources of statistics, periodicals, abstracting services, and films as well as books. The Library of the University of Akron, in association with the ACS Rubber Division, produces bibliographies on specific subjects: these are listed in *Rubber Chemistry and Technology*. This source also used to compile the *Bibliographies of Rubber Literature* (annual, 1935–1972), and the Librarian (Ruth Murray) was a co-author of the annotated bibliography contained as an appendix to Morton (1987). The annual bibliography was compiled from *Chemical Abstracts* and *RAPRA Abstracts*: the standard of indexing was high.

Dictionaries

The rich terminology of rubber technology is described by Byam (1968). For definitions, there is a choice between compilations based upon working lifetimes by Craig (1969) and Heinisch (1974, 1977), and the products of standardizing committees (ASTM, BSI and ISO). Craig is illustrated with diagrams. Heinisch contains many common abbreviations.

The BSI (1980) glossary is evaluative and incorporates what it calls deprecated terms. Diagrams are used to relate terminology to components. BSI (1978) also publishes a list of abbreviations for the major rubbers. The ISO (1982) glossary also contains French and Russian definitions: some of which are excessively long. The glossary produced by ASTM (1972) contains lengthy definitions of about 1500 terms. There are some errors in cross-referencing.

Two aged polyglot dictionaries omit many common expressions whilst including many dubious translations: Dorian (1965) and Elsevier (1959) are based on English and contain translations in Dutch, French, German, Italian and Spanish, whilst Elsevier also includes Indonesian, Japanese (Nihongo), Portuguese and Swedish. There is a much shorter, but more thorough polyglot glossary based on Swedish (Swedish Centre — 1968) with English, French and German translations. In all cases there are indexes for the non-chosen languages.

Simpler translating dictionaries are less common. There is a miniscule, but very useful French-, German- and Spanish-English vocabulary produced by Imperial Chemical Industries plc, a Polish-English-Polish dictionary compiled by Dziunikowskiego (1965) and a Russian-English dictionary compiled by Lambert (1963). This last includes extensive annotations.

Directories

The most extensive directory was the *International Rubber Directory* (last published 1975, Verlag für Internationale Wirtschaftsliteratur) which appears to have ceased publication. It covered all continents (the primary division) and was then subdivided by countries, towns and names of firms: companies located in suburbs are difficult to trace. Entries may be in German, French or English. All entries quote address, telephone numbers, number of employees and product ranges. Trade and research associations are listed separately. Company names, but not products, are indexed.

The *Rubber Red Book* (annual, Communications Channels) covers the USA and Canada. The entries for product manufacturers are detailed. Normal directory information is augmented by date of establishment, employee numbers, senior management, plant locations and products. Broad headings are used to list products which makes reference difficult. Factories are indexed geographically. Suppliers are listed with less detail, and there are listings of processing machinery, laboratory equipment, rubber chemicals, trade and brand names, textiles, synthetic rubbers, natural rubber, consultants, educational institutions, trade and technical organizations and technical journals. The *Blue Book* (annual, Lippincott and Peto) is the major source of information about rubber-compounding ingredients.

Rubbicana (annual, Crain Automotive Group) commenced publication after the *Rubber Red Book* and in the case of the North American edition covers similar ground, but in greater detail for the product manufacturers. Additional information includes gross sales, key personnel, markets served, base rubbers used, and equipment. It also suffers from the broadness of the headings used for products and includes suppliers, rubber types and properties and consultants. There is what purports to be a who's who, but is an index to personnel; nevertheless this is useful in the absence of a biographical directory.

Rubbicana Europe (biannual, Crain Communications) is new and fills a major gap in the European literature. It is similar to the North American edition listing for product manufacturers: address, telephone and telex numbers, number of employees, year founded, key personnel, factory locations, products, subsidiaries, sales offices and materials used. The headings to the classified lists are accompanied by French, German and Italian translations. There are some serious omissions, notably Michelin. In contrast, the North American edition records companies which fail to return questionnaires. Avon is not credited with tyre manufacture in the indexes.

The *BRMA Directory: rubber and polyurethane* (irregular, latest 1989, British Rubber Manufacturers' Association) is well organized using the matrix system associated with Kompass. Products are defined very precisely within sensible broad groupings. The headings are translated into French, German and Spanish. There are excellent alphabetical indexes.

The *Scandinavian Rubber Handbook* (Nodisk Gummiteknisk Handbok) (irregular latest 1983, Sveriges Gummitekniska Förening) lists producers (addresses, telephone numbers and factories only) and organizations associated with the industry (including some non-Scandinavian). It is an excellent source of information about compounding ingredients and processing machinery as there is a thorough index of tradenames. The material is mainly in Swedish, but English headings are provided. The *Guide du Syndicat général des Commerces et Industries du Caoutchouc et des Plastiques* (annual, Aprocap) lists French manufacturers: there is an English index of products. *Rubber Industry in Japan* (annual, Rubber Times, Tokyo) includes a directory of rubber manufacturers and detailed listings of products available.

Abstracting services

RAPRA Abstracts is the major abstracting service. It covers a very extensive period (back to 1923), but lacks subject indexes prior to 1947. The abstracts are available online back to 1972 via Infoline. The online service may be searched via free-text, controlled terminology, or RAPRA classification codes: a guide is available. Coverage now tends to be restricted to a core set of journals: patents are not abstracted. For wider coverage it is necessary to search *Chemical Abstracts, Physics Abstracts* and the services provided via the CAB International: all are available via several online hosts.

Journal literature

Rubber Chemistry and Technology (five issues per annum, ACS Rubber Divsion) holds a pivotal position in the rubber literature. One issue (entitled *Rubber Reviews*) is dedicated to major review articles with several hundred citations. Other sources of review articles are *Progress in Rubber and Plastics Technology* (quarterly, RAPRA Technology) and the monograph series *Developments in Rubber Technology* (irregular, Applied Science). *Rubber Chemistry and Technology* publishes abstracts of papers presented at meetings of the ACS Rubber Division, original papers and reprints of material published elsewhere. The journal used to publish translations of non-English language papers. There is an excellent annual index with ten-year cumulations.

Rubber and Plastics News (Crain Communications), alternating with *Rubber and Plastics News II*, together provide a weekly source of news. The same publisher is responsible for *Automotive News* (weekly), *Tire Business* (fortnightly) and *European Rubber Journal* (monthly, except August). There is considerable overlap between the American and European publications, but less with the motor industry journal. *Plastics and Rubber Weekly* (Maclaren) used to serve a similar function, but is now dominated by news about plastics.

European Rubber Journal is characterized by its investigative journalism and interest in business affairs. Extensive data on items like tyre plant capacities may be included. There are occasional biographical features. Both *Elastomers* (monthly, Communications Channels) and *Rubber World* (monthly, Hartman Communications) have a more technical bias, especially the first-named. Both contain some news, a calendar, biographical notes and book reviews. *Rubber World* contains lists of raw material prices. *Kautschuk und Gummi Kunststoffe* (monthly, Hüthig) is the journal of the DKG. It contains news, technical articles (some of which are in English), reports of other conferences and abstracts of technical literature. *Industria della Gomma* (monthly, ASSOGOMMA) contains news, technical and techno-economic articles and statistics. The articles include an English translation of the title. *Revue générale de Caoutchouc et Plastiques* (nine issues per annum, Société d'Expansion Technique et Economique) is the French equivalent of the German and Italian journals, but also includes thermoplastics.

Conferences

The premier event is the annual International Rubber Conference which may be organized under other titles, such as Rubbercon in the United Kingdom. The proceedings may be well presented, or consist merely of bound copies of preprints. Sometimes they are difficult to trace. The PRI is responsible for the British event.

Other major conferences occur twice per annum under the auspices of the ACS Rubber Division (abstracts are published in *Rubber Chemistry and Technology*), the DKG (papers eventually may appear in *Kautschuk und Gummi Kunststoffe* and the Scandinavian Rubber Conference (publication may be organized in any of the Scandinavian countries).

Economics

Rubber is well served with literature on its economics and with statistical information. Monographs include: Allen (1972) which is now in need of revision; Smit (1982), a major econometric study; Barlow (1978) and Grilli (1980) on natural rubber (the latter being a prestigious World Bank study) and Arlie (1980) on synthetic rubbers.

The publications of the International Rubber Study Group are the cornerstone for statistical data. There are three volumes of the *World Rubber Statistics Handbook*: 1946–1970, 1965–1980 and one which extends the range of data. This last is available in machine-readable form on floppy discs. *Rubber Statistical Bulletin* provides monthly data with annual cumulations of production, consumption, stocks and trade broken down by types of rubber and countries. Currently it contains 46 tables. There are data for tyre production and rubber consumption in tyres for several major countries. The *Proceedings of the International Rubber Study Group* (annual or semi-annual), especially the associated Discussion Forums (which may be published separately) contain a wealth of data on specific topics.

Rubber Trends (quarterly, Economist Intelligence Unit) tends to discuss one or two specific topics in each issue: an issue may include a survey of one country in terms of its natural rubber production or its rubber consumption and/or the economic health of one large manufacturer such as Goodyear. There is also a short section of statistical news and some regular tables relating to the position of the industry in major producing and consuming nations. Landell Mills Commodities Studies Ltd. produces both monthly and quarterly bulletins which are primarily intended for the investment community. They consist of a mixture of highly informed comment, data (including regular rubber consumption indicators for the USA, Japan, the EEC and Comecon), and in the case of the quarterly, projections. Considered as publications these may appear to be expensive, but they are highly valuable aids to investment decisions.

Rubber Industry in Japan (annual, Rubber Times, Tokyo) is an English language annual reveiw of the Japanese rubber industry, which includes detailed time-series of statistical data in greater depth than is normally found. Thus there are statistics on the production of fenders and vibration isolators and on the end-uses of rubberized fabrics.

Specialized forms

Latices are discussed by Calvert (1982) and his co-authors, but they receive more comprehensive treatment from Blackley (1966) and Noble (1953). There is no specific journal, but MRPRA produces a series of *Natural Rubber Information Sheets*. The PRI has organized two Latex Conferences. Powdered rubber is examined with excessive optimism by Evans (1978). Thermoplastic rubbers are thoroughly reviewed by Legge and co-editors (1987). The "comprehensive review" of the sub-title is a justifiable claim.

Natural rubber

The botany of most rubber-bearing species is covered in Polhamus (1962); *Hevea brasiliensis* receives extended treatment, it can be augmented by Compagnon (1986), which also examines agricultural and plantation aspects and by Auzac (1988). Rubber cultivation is covered in depth by Pee and Ani (1976) which has largely superceded the frequently sought Edgar (1958). The science, technology, engineering and innovation within the natural rubber industry were reviewed by Roberts and co-authors (1988). Guayule is the only other potential, commercial source of natural rubber: National Academy (1977) provides an introduction.

The RRIM is the major centre for research into the production of natural rubber, the literature has been recorded by Soosai and Kaw (1975). It also publishes the *Directory of Registered SMR Producers in Malaysia* (irregular, latest 1985). The *Journal of Natural Rubber Research* (quarterly, RRIM) has grown from the *Journal of the Rubber Research Institute of Malaysia*, but is now broader in scope, by including consumption as well as production related research. *Planters' Bulletin* (quarterly, RRIM) contains advice on agricultural activities. *NR Technology* (quarterly, MRPRA) has a self-evident slant: it also includes abstracts of recent MRPRA publications. *Rubber Developments* (quarterly, MRPRA) is aimed at a wide readership to promote natural rubber usage. English language journals are also published by the Rubber Research Institute of Sri Lanka and the Rubber Board of India. The Annual Report of the IRRDB provides an excellent short-cut to current research in major producing territories.

Malaysian Rubber Review (quarterly, MRRDB) contains statistical material. Most of the data relate to Malaysia; there is a breakdown of local consumption by sectors. Natural rubber prices are monitored: comparison of the Kuala Lumpur, London and New York prices is assisted by a table of exchange rates.

The *RRIM Statistical Bulletin* enumerates shipments of Standard Malaysian Rubber by categories monthly and shipments of speciality rubbers quarterly. The *Monthly Bulletin of the Malaysian Rubber Producers' Council* may be used to augment these data.

MRRDB and its units are responsible for a range of other literature: the *MRRDB Monographs* examine the production, economics and marketing of Malaysian natural rubber, *Natural*

Rubber Technical Information Sheets (in two series — one for latex and one for dry rubber), *Natural Rubber Engineering Data Sheets* and *Natural Rubber Technical Bulletins*. Some of this could be described as trade literature, but trade literature *per se* is relatively uncommon in the natural rubber industry. MRRDB is also responsible for several conferences, notably the annual *Rubber Growers' Conference* which since 1986 has taken the place of the *Planters' Conference*.

Synthetic rubber

The IISRP (1973) produced an elegant, well illustrated guide to the industry and Arlie (1980) has examined its economics in depth. Kaufman (1983) highly commends Blackley (1983) for his textbook on the science of synthetic rubbers, but notes that their technology (covered more fully by Blow and Hepburn (1982), Morton (1987) and Brydson (1988)) is lightly treated. Kennedy and Tornqvist (1968–1969) is also mainly concerned with polymerisation. There is a surprising paucity of monograph literature on specific synthetic rubber types, the following are exceptions, however: Saltman (1977) on stereo-rubbers, Lynch (1978) on silicone rubber and Hepburn (1982) on polyurethane elastomers. There is the much broader polymer literature which examines them. The *Encyclopedia of polymer science and technology* (Chapter 5) contains many pertinent articles.

In contrast with natural rubber there are few journals devoted to synthetic rubber. This is partly because the science is closely allied to other polymer science and partly because much of the synthetic rubber output has a captive market: most major tyre manufacturers possess synthetic rubber plants. Most of the more specialised elastomers are traded and do support a periodical trade literature, notably the Du Pont *Elastomers Notebook* (which covers polychloroprene, fluoroelastomers and chlorosulphonated polyethylene amongst others). Du Pont also produce loose-leaf material on elastomers are traded and do support a periodical trade literature, literature for their extensive materials which include compounding ingredients as well as elastomers. The Bayer (1972) manual is a particularly useful compendium and this is augmented by the irregular *Technical Notes for the Rubber Industry* and *Technical Information Bulletins* (which are actually data sheets) for each major group of materials, such as polychloroprene, Shell, Exxon and Polysar also produce a considerable amount of, mainly loose-leaf, literature. Shell has a separate *Technical Manual* for its range of thermoplastic rubbers.

Worldwide Rubber Statistics (annual, IISRP) incorporates time series on production (by area and by type of elastomer), supply, consumption and per capita consumption, synthetic rubber plant capacities (by country and then by actual facilities), key vehicle statistics, and forecasts of future rubber production. The data are presented in tabulated and graphical forms. There is a directory of producers. This is augmented by the *Synthetic Rubber Manual* (irregular, latest 1986, IISRP) which provides a detailed breakdown of the types of synthetic rubber available and their manufacturers. In addition the *Proceedings* of the annual meeting of the IISRP provide a source for further statistics and informed comment about the state of the industry. The International Plastics Selector (1980) is a useful compendium of mainly pre-compounded materials and standard formulations.

Compounding ingredients

The *Blue Book* (annual, Lippincott and Peto) is the major source of information about rubber-compounding ingredients, especially those supplied within the USA. The primary division is by type of ingredient and then by tradename. Information includes composition, supplier, properties and function. There is data about rubber types and some information on processing machinery. The directory is well indexed. Further data may be found in Van Alphen (1973); Mal (1983); Vanderbilt (1978); Bayer (1972); the British Rubber Manufacturers' Association (1978) Code of Practice for safe-handling; in books on accelerators by Blokh (1978) and carbon black by Donnet and Voet (1976), and in the extensive trade literature.

Tyres

Tyres form the major outlet for rubber: over 60% of both natural rubber and styrene butadiene rubber are consumed in them. Some major developments in tyre technology are protected by patents, but much expertise is maintained by secrecy, and if it is transferred then it is protected through know-how agreements. Thus the literature on tyres is limited. Setright (1972) is one of the few popular accounts of tyre technology. Kovac (1973) is difficult to characterize as it is partly technological (with chapters on testing and reinforcement) and partly on economics. It includes a glossary, a chronology and statistics of USA tyre production back to 1910. Clark (1982), Hays and Browne (1974), Moore (1975) on friction and Fleming and Livingstone (1979) represent more specialised contributions. The history of tyres is presented with a slant towards Dunlop innovations by Tompkins (1981).

Tire Science and Technology (quarterly, Tire Society, Akron) started publication in 1973, but struggled to survive until 1985, since when production has been regular. *Modern Tire Dealer* (monthly — semi-monthly in April and September, Bill Communications) is a major source of commercial information. The January issue usually contains an extensive statistical survey of the USA market. A special product review issue is published in April. *Tire Business* (forthnightly, Crain Communications) covers similar ground. *TAB* (bimonthly, Municipal Publications) and *Tyres and Accessories* (monthly, Tyre Industry Publications) are the British equivalents, but lack the depth of coverage of their USA counterparts.

The Smithers Laboratories produce a wide range of specialised analytical and market surveys of tyres on the US, European and Japanese output.

Other end uses

There are many other uses for rubbers; some, like adhesives and footwear have distinctive literatures of which the elastomeric element forms a part. Products more closely identifiable with the rubber industry include hose which is discussed by Evans (1974), belting by Conveyor (1979), and seals by Flitney (1984) and Warring (1967). Rubber has been used as an engineering material for over a century; see Woodruff (1958). Design criteria are set forth in Lindley (1974) and Freakley and Payne (1978). The former is extremely concise and is restricted to natural rubber, although the same author used to be responsible for the chapter on rubber in *Kempe's Engineers' Yearbook* which includes a wide range of rubbers. Freakley and Payne was in effect the third edition of a book originally written by Payne with other co-authors: Lindley (1979a) described it as 'an excellent reference book for engineers'.

Case studies, plus some theory, are incorporated in Hepburn and Reynolds (1979) which Lindley (1979b) described as a 'fitting and permanent tribute to Bob Payne'. Books on specific sectors of engineering include Derham (1983) on the design of earthquake-resistant buildings mounted on rubber springs and Stevenson (1984) on off-shore engineering.

References

Allen, P.W. (1972). *Natural rubber and the synthetics* (Granada).

Alliger, G. and Sjothun, I.J. (1964). *Vulcanisation of elastomers* (Reinhold).

American Society for Testing and Materials (1972). *Glossary of terms relating to rubber and rubber technology* (STP 184A).

Arlie, J.P. (1980). *Caoutchoucs synthetiques: procedes et données economiques* (Editions Technip).

Auzac, J. d', Jacob, J.L. and Chrestin, H. (1988). *Physiology of rubber tree latex* (CRC).

Barlow, C. (1978). *The natural rubber industry* (OUP, Kuala Lumpur).

Barlow, F.W. (1988). *Rubber compounding* (Dekker).

Barnard, D. (1979). [Review]. *Rubber Developments*, **32**, 21.

Bateman, L. (1963). *The chemistry and physics of rubber-like substances* (Maclaren).

Bayer (1972). *Manual for the rubber industry* (Farbenfabriken Bayer).

Berry, J.P. and Stanford, J.L. (1987). [Obituary]. *Polymer*, **28**, 355.

Blackley, D.C. (1966). *High polymer latices* (Maclaren).

Blackley, D.C. (1983). *Synthetic rubbers: their chemistry and technology* (Applied Science).

Blokh, G.A. (1978). *Organic accelerators and curing systems for elastomers*; translated R.J. Moseley (RAPRA).

Blow, C.M. and Hepburn, C. (1982). *Rubber technology and manufacture* (2nd edn, Butterworth).

British Rubber Manufacturers' Association. (1978). *Toxicity and safe handling of rubber chemicals.*

British Standards Institution. (1978). *Schedule of common names and abbreviations for plastics and rubbers. Part 3.*

British Standards Institution. (1980).*Glossary of rubber terms.*

Brown, R.P. (1979a). *Guide to rubber and plastics test equipment.* (RAPRA).

Brown, R.P. (1986). *Physical testing of rubber* (2nd edn, Elsevier).

Brydson, J.A. (1978). *Rubber chemistry* (Applied Science).

Brydson, J.A. (1988). *Rubbery materials and their compounds* (Elsevier).

Byam, S.C. (1968). 'The jargon of the rubber industry'. *Advances in Chemistry Series*, **78**, 429.

Calvert, K.O. (1982). *Polymer latices and their applications* (Applied Science).

Clark, S.K. (1982). *Mechanics of pneumatic tires* (US Department of Transportation).

Coates, A. (1987). *The commerce in rubber* (OUP, Singapore).

Compagnon, P. (1986). *Le caoutchouc naturel: biologie; culture; production* (Maisonneuve and Larose, Paris)

Conveyor Equipment Manufacturers Association. (1979). *Belt conveyors for bulk materials* (2nd edn).

Craig, A.S. (1969). *Dictionary of rubber technology.* (Newnes-Butterworths).

Dean, W. (1987). *Brazil and the struggle for rubber* (CUP).

Derham, C.J. (1983). *Proceedings of the International Conference on Natural Rubber for Earthquake Protection of Buildings and Vibration Isolation* (MRRDB).

Donnet, J.B. and Voet, A. (1976). *Carbon black: physics, chemistry and elastomer reinforcement* (Dekker)

Dorian, A.F. (1965). *Six-language dictionary of plastics and rubber technology* (Iliffe).

Drabble, J.H. (1973). *Rubber in Malaya, 1876–1922* (OUP, Kuala Lumpur).

Dziunikowskiego, W. (1965). *The little English-Polish and Polish-English dictionary* (Biuro Projektow Przemysku Gumowego).

Edgar, A.T. (1958). *Manual of rubber planting* (Incorporated Society of Planters,

Kuala Lumpur).
Eirich, F.R. (1978). *Science and technology of rubber* (Academic).
Elsevier. (1959). *Rubber dictionary in ten languages.*
Evans, C.W. (1974). *Hose technology* (Applied Science).
Evans, C.W. (1978). *Powdered and particulate rubber technology* (Applied Science).
Evans, C.W. (1981). *Practical rubber compounding and processing* (Applied Science).
Fleming, R.A. and Livingston, D.I. (1979). *Tire reinforcement and tire performance* (ASTM).
Flitney, R.K., Nau, B.S. and Reddy, D. (1984). *The seal users handbook* (3rd edn, BHRA).
Freakley, P.K. and Payne, A.R. (1978). *Theory and practice of engineering with rubber* (Applied Science).
Freakley, P.K. (1985). *Rubber processing and production organisation* (Plenum).
Genin, G. and Morisson, B. (1956–1960). *Encyclopedie technologique de l'industrie du caoutchouc* (Dunod).
Goodyear, C. (1855). *Gum-elastic* (Author, New Haven).
Grilli, E.R., Agostini, B.B. and Hooft-Welvaars, M.J. (1980). *The world rubber economy* (Johns Hopkins University Press).
Hancock, T. (1857). *Personal narrative* (Longman).
Hays, D.F. and Browne, A.L. (1974). *The physics of tire traction* (Plenum).
Heinisch, K.F. (1974). *Dictionary of rubber* (Applied Science).
Heinisch, K.F. (1977). *Kautschuk-lexicon* (Gentner).
Hepburn, C. and Reynolds, R.J.W. (1979). *Elastomers: criteria for engineering design* (Applied Science).
Hepburn, C. (1982). *Polyurethane elastomers* (Applied Science).
Herbert, V. and Bisio, A. (1985). *Synthetic rubber* (Greenwood).
Hills, D.A. (1971). *Heat transfer and vulcanisation of rubber* (Elsevier).
Hofmann, W. (1967). *Vulcanisation and vulcanizing agents* (Maclaren).
International Institute of Synthetic Rubber Producers. (1973). *Synthetic rubber.*
International Organisation for Standardisation. (1982). *Rubber vocabulary* (ISO 1382).
International Plastics Selector. (1980). *Elastomers: desk-top data bank* (Cordura).
Kaufman, M. (1983). [Reviews]. *Chemistry and Industry*, 282.
Kennedy, J.P. and Tornqvist, E.G.M. (1968–1969). *Polymer chemistry of synthetic elastomers* (Interscience).
Kirkham, M.C. (1978). 'Current status of elastomer vulcanisation'. *Progress of Rubber Technology*, **41**, 61–95.
Kovac, F.J. (1973). *Tire technology* (Goodyear, Akron).
Lambert, M. (1963). *A short Russian-English dictionary of terminology* (Maclaren).
Le Bras, J. (1968). *Introduction to rubber* (Applied Science).
Legge, N.R., Holden, G. and Schroeder, H.E. (1987). *Thermoplastic elastomers* (Hanser).
Lindley, P.B. (1974). *Engineering design with natural rubber* (4th edn, MRPRA).
Lindley, P.B. (1979a). [Review]. *Plastics Rubber International*, **4**, 87.
Lindley, P.B. (1979b). [Review]. *Plastics Rubber International*, **4**, 191.
Long, H. (1985). *Basic compounding and processing of rubber* (ACS Rubber Division).
Lynch, W. (1978). *Handbook of silicone rubber fabrication* (Van Nostrand).
Mal, G. (1983). *Dictionary of rubber chemicals* (Modi Rubber, India).
Mark, J.E. and Lal, J. (1982). *Elastomers and rubber elasticity* (ACS).
Martin Sweets Co. (1971). *A glossary of urethane industry terms.*
McMillan, F.M. (1979). *The chain straighteners* (Macmillan).

Moore, D.F. (1975). *The friction of pneumatic tyres* (Elsevier).
Morton, M. (1987). *Rubber technology* (3rd edn, Van Nostrand).
National Academy of Sciences. (1977). *Guayule: an alternative source of natural rubber*.
Naunton, W.J.S. (1961). *The applied science of rubber* (Arnold). Noble, R.J. (1953). *Latex in industry* (Rubber Age).
Pee, T.Y. and Ani bin Arope. (1976). *Rubber owners' manual* (RRIM).
Penn, W.S. (1969). *Injection moulding of elastomers* (Maclaren).
Phillips, C.F. (1963). *Competition in the synthetic rubber industry* (University of North Carolina).
Polhamus, L.G. (1962). *Rubber: botany, production and utilisation* (Leonard Hill).
Porter, M. (1979). [Review]. *Rubber developments*.
Roberts, A.D. (1988). *Natural rubber science and technology* (OUP).
Roff, J. and Scott, J.R. (1971). *Fibres, films, plastics and rubbers* (Butterworths).
Saltman, W.M. (1977). *The stereo rubbers* (Wiley).
Schidrowitz, P. and Dawson, T.R. (1952). *History of the rubber industry* (Heffer).
Setright, L.J.K. (1972). *Automobile tyres* (Chapman and Hall).
Seymour, R.B. (1982). *History of polymer science and technology* (Dekker).
Small Business Publications. (1983). *Rubber technology and manufacture* (Delhi).
Smit, H.P. (1982). *The world rubber economy to the year 2000* (Free University of Amsterdam).
Smith, J.F. (1979). [Review]. *Plastics and Rubber International*, **4**, 142–3.
Soosai, J.S. and Kaw, H.W. (1975). *Fifty years of natural rubber research: 1926–1975* (RRIM).
Sowry, J. (1977). *Rubber* (Priory). Stern, H.J. (1967). *Rubber: natural and synthetic* (2nd edn, Maclaren).
Stevenson, A. (1984). *Rubber in offshore engineering* (Hilger).
Swedish Centre of Technical Terminology. (1968). *Glossary of rubber terms: Swedish-English-German-French*.
Tompkins, E. (1981). *The history of the pneumatic tyre* (Dunlop).
Treloar, L.R.G. (1975). *The physics of rubber elasticity* (3rd edn, OUP).
UNIDO. (1979). *Information sources on the natural and synthetic rubber industry*.
Van Alphen, J. (1973). *Rubber chemicals* (Reidel).
Vanderbilt (1978). *The Vanderbilt rubber handbook*.
Wake, W.C. and Wootton, D.B. (1982). *Textile reinforcement of elastomers* (Applied Science).
Wake, W.C., Tidd, B.K. and Loadman, M.J.R. (1983). *Analysis of rubber and rubber-like polymers* (3rd edn, Applied Science).
Warring, R.W. (1967). *Seals and packings* (Trade and Technical).
Wheelans, M.A. (1974). *Injection moulding of rubber* (Butterworths).
Whitby, G.S., Davis, C.C. and Dunbrook, R.F. (1954). *Synthetic rubber* (Wiley).
Wickham, H.A. (1908). *On the plantation, cultivation and curing of PARA Indian rubber* (Kegan Paul).
Woodruff, W. (1958). *The rise of the British rubber industry during the nineteenth century* (Liverpool University).
Yescombe, E.R. (1968). *Sources of information on the rubber, plastics and allied industries* (Pergamon).
Yescombe, E.R. (1976). *Plastics and rubber: world sources of information* (Applied Science).

CHAPTER FIFTEEN

Coatings and adhesives

S.C. HAWORTH

Introduction

From inside these industries, it is tempting to say that the chief difficulty is that almost every subject is, or will sometimes be, relevant. Quite apart from the composition, formulation, manufacture, properties and application of these substances they cannot be considered without also considering the materials with which they are used and to which they must adhere. This may lead one, for example, to the study of steel or of teeth, for both of which coatings and adhesives are used.

Polymer science is of great importance, as are all the aspects of chemistry and physics involved. Synthetic chemistry, analytical chemistry, chemical engineering, lead on to the physical sciences and technologies: colour science, rheology, adhesion, other mechanical properties; and to biology and microbiology: microbial degradation of paint in cans, biological growth on paint films, fouling on underwater structures. The substrates whose nature must be understood may consist of almost anything: metals, especially steel, plastics, and inorganic building materials are of major importance for paints; materials such as paper are also of importance for adhesives. This introduces metallurgy, corrosion science, different topics in polymer chemistry and technology, paper science. Another field of great importance to everyone is that of the associated hazards, toxicology and pollution, for which the literature is growing at an alarming rate, as described by O'Neill (1981 and 1986) and in *Survey of Hazards, Pollution and Legislation in the Coatings Field* (quarterly, Paint Research

Association). Yet another field which cannot be ignored is that of commercial information: company data, statistics and market news, market trends and developments. Moreover, many associate paint with paintings and this could introduce the questions of art, art history, aesthetics, historical materials and so ad infinitum.

This multiplicity of subjects inevitably leads to a wide range of sources of information, from very general publications to very specialised ones, from all over the world. Several hundred journals will supply the majority of the useful information but many more will sometimes supply important data. As with all scientific and technological subjects, most of the important information is published first in the journals, or else in conference papers, patents, standards, and the textbooks have the same role as review articles in journals, providing a survey of information available at the time of writing.

Abstract journals

The two principal organizations concerned with providing specialist information, and themselves publishing abstract journals, are the Paint Research Association (Paint R.A.) and RAPRA Technology Ltd. Both are research associations established by and for the UK industries and have since broadened their scope to extend membership to companies from many other countries, and to undertake research and supply information for anyone interested, on appropriate terms. Both take, and abstract, several hundred primary journals, and the Paint R.A. also scans other abstract journals, to extend its coverage of the literature.

The central source of information on paints, and also on adhesives, is *World Surface Coatings Abstracts (WSCA)* (monthly, Pergamon Press for Paint R.A.) published in hard copy with annual printed subject and author indexes and also available online on Pergamon Orbit Infoline. Each year it contains about 10,000 abstracts and the computer-held file contains more than 100,000 abstracts. About 40% of the abstracts are summaries of patents and the remainder are of journal articles, conference papers, standards, reports, as well as book reviews. The subjects covered include paint raw materials (pigments, resins, additives), many different types of coating, adhesives, inks, manufacturing plant and processes, pretreatment and application methods, paint removal, weathering, corrosion, fouling, microbiological attack, optical properties and colour science and technology, mechanical and rheological properties, analytical methods, industrial and other hazards, pollution and utilisation of wastes, packaging, storage and transport, radiation curing, economics, statistics, market news, organisation news, education, training and information, energy, legislation and other official publications, standards and specifications, book reviews. Adhesives are also covered in *RAPRA Abstracts* (monthly, Pergamon Press for RAPRA Technology Ltd.). As with *WSCA*, this is available in hard copy and online, and contains about 2400 abstracts each year, with printed subject index. *RAPRA Abstracts* covers journals, conference proceedings, books and company literature, with sections for commercial and economic information, legislation, health and safety, raw materials and monomers, polymers and polymerisation, compounding ingredients, applications (including coatings), processing and treatment, properties and testing. Another abstract journal which includes both inorganic and organic coatings and surface treatments is *Surface Treatment Technology Abstracts* (bimonthly, Finishing Publications) which concentrates on metal protection. From 1986 onwards RAPRA will be publishing a separate abstract journal, *Adhesives Abstracts* based on *RAPRA Abstracts* and also available online, covering the literature on adhesives and adhesion. Other abstract journals including adhesives among other subjects are *International Packaging Abstracts* (monthly, Pergamon Press for PIRA), and the *Abstracts Bulletin of the Institute of Paper Chemistry* (monthly, Institute of Paper Chemistry).

Another secondary source of information on published literature about both paints and adhesives is *Paint Titles* (weekly, Pergamon Press for Paint R.A.) which lists in subject-classified order the titles, expanded where necessary to make the meaning clear, of newly received journal articles, conference papers, technical literature, etc. and also contains summaries of new European and British patents. This caters for those who need or prefer to scan rapidly across a wide subject field, or to see what has just been published, and is used by some organizations instead of a circulation system for their journals.

Apart from these 'dedicated' abstract journals, many others may also be useful on particular topics, especially the myriad associated subjects, most available both as hard copy and online.

Primary journals

The most important scientific and technical journals on coatings are *Journal of Coatings Technology* (monthly, Federation of Societies for Coatings Technology), *Journal of Protective Coatings and Linings* (monthly, Steel Structures Painting Council), *JOCCA* (monthly, Oil and Colour Chemists' Association (OCCA)), *Farbe und Lack* (monthly, Curt R. Vincentz Verlag), all of which contain substantial articles on specific topics, as well as industrial news. Other journals with varying proportions of technical articles, industrial and commercial news, press releases and disguised advertisements, include *Surface Coatings Australia* (monthly, Oil and Colour Chemists' Association Australia), *American Paint and Coatings Journal* (weekly, American Paint Journal Co.), *Double Liaison* (monthly, A.F.T.P.V.), *Finishing* (monthly, Turret-Wheatland Ltd.), *Industrie Lackierbetrieb* (monthly, Curt R.Vincentz Verlag), *Modern Paint and Coatings* (monthly, Communication Channels Inc.), *Paint and Resin* (bimonthly, Turret-Wheatland Ltd.), *Pigment and Resin Technology* (monthly, Sawell Publications Ltd.), *Polymers Paint Colour Journal* (fortnightly, Fuel and Metallurgical Journals Ltd.), *Product Finishing* (monthly, Sawell Publications Ltd.).

Journals specifically on adhesives include *Adhäsion* (monthly, Verlag Heinrich Vogel Fachzeitschriften GmbH), *European Adhesives and Sealants* (quarterly, Fuel and Metallurgical Journals Ltd.), *International Journal of Adhesion and Adhesives* (quarterly, Butterworth Scientific Ltd.), *Journal of Adhesion* (irregular, Gordon and Breach).

These two lists are not mutually exclusive: articles on coating and related subjects occur in the 'adhesives' journals, and articles on adhesives in the 'coatings' journals. Two useful lists of journals covering both coatings and adhesives and also the many related topics are those in *WSCA* and in *RAPRA Abstracts*, listing the journals covered by the abstract journals. As is particularly clear in the list of journals scanned for *WSCA*, useful literature comes from all over the world. While the majority of the journals are in English and many others have English-language summaries of the articles, there is much of importance published in German and in many other languages.

Dictionaries

As well as general dictionaries for particular pairs of languages, there are those concentrating on the technical terms peculiar to individual industries. Two multi-lingual dictionaries of specialised coverage are those by Santholzer (1969) offering English-American, Czech, Russian, German and French terms for surface coatings, finishing, corrosion, plastics and rubber and that by

Raaff (1965) providing Dutch, French, English and German equivalent words for paints. Another multi-lingual dictionary is the Swedish glossary of paint terms, *Tekniska nomenklatur-centralens* (1967), offering Swedish, English, French, German and Finnish, with a definition of every term in Swedish. Yet another is Clausen (1980) who provides a dictionary of paint technology in German, English and French. There are two useful dictionaries for German and English words: Ernst (1974) provides general coverage of 'industrial techniques' and Gross (1986) for corrosion and corrosion control terms. For French and English the Association Quebécoise des Industries de la Peinture (1977) has published a small but useful dictionary for the paint and coatings industry. There are also a number of multi-lingual dictionaries for particular industries published by Elsevier, e.g. Dettner (1971) giving English, French, Italian, Dutch and German words for metal finishing and corrosion.

Review articles

The prime source of review articles is *Progress in Organic Coatings* (quarterly, Elsevier) but useful review articles may appear in many other journals and every other year the review issue of *Analytical Chemistry* (fortnightly, American Chemical Society) contains a review article surveying the previous two years' publications on the analysis of coatings. A list of review articles was published as *Information Brief no. 1* by the Paint R.A. in 1984 and they are abstracted in *WSCA, RAPRA Abstracts*, etc.

Books

As in so many technical fields, the importance of books is reduced by the fact that the information they contain is already obsolescent at the time of publication. In the case of general books on adhesives and particularly on paints the problem is exacerbated by the wide diversity of subjects to be covered within a single book. A basic introductory book on paint technology is published by OCCA (1976) and updated at intervals. A more detailed, multi-author book covering both constituents and types of paints is a two-volume work from OCCA (1983, 1984). A useful book covering surface preparation and specific functional types of paints is Weismantel (1981) with chapters written by different authors, a clear layout and a good index. Another general textbook covering both raw materials and finished products is that by Morgans (1982, 1984). The range of more specialised works on specific topics is large, but a useful start is the introduction to paint chemistry and principles of paint technology by Turner (1980). For analytical chemistry, two basic sources are Hummel/Scholl (1978–1984) on analysis of polymers and plastics, containing many spectra, and the infrared spectroscopy atlas from the Chicago Society for Coatings Technology (1980). Colour science concepts and methods are dealt with extensively by Wyszecki and Stiles 1982). For pigments the major reference works are the *Pigment Handbook* (Patton, 1973) with chapters on individual pigments, applications and characterisation methods, and the *Pigments* and *Solvent Dyes* volume of the *Colour Index* (Society of Dyers and Colourists, 1982) for data. Solomon and Hawthorne (1983) cover the chemistry of pigments and fillers and Parfitt (1973) and Patton (1979) deals with the incorporation of pigments into paints and their rheology. On resins for paints, and specific chemical types of paints, two American Chemical Society books (1975 and 1985) provide many review articles and also cover the chemistry and physics of paints and adhesives.

Books on specific types of paints and their application include Munger (1984) for high-performance, anti-corrosive coatings, British Maritime Technology (1986) for painting of ships, Hurst and Goodier (1980) for painting and decorating, Satas (1986) for plastics finishing and decorating, Roffey (1982) for photopolymerizable coatings, Hughes (1984) for powder coatings, Holman (1984) for ultraviolet and electron beam curing formulations. The classic work on paint film defects is Hess (1979).

A book relevant both to adhesives and to coatings is Mittal (1983), papers from a symposium on the adhesion aspects of polymeric coatings with much theoretical consideration of the phenomena involved. A less theoretical multi-author book on the adhesion of adhesives and of coatings is by Brewis and Briggs (1985). Lee (1984), again with much theory, has edited papers on adhesive chemistry developments and trends.

Two handbooks covering different types of adhesives and their application are Shields (1984) and Skeist (1977), and a useful general textbook on adhesion and the formulation of adhesives is that by Wake (1982).

This section could be prolonged indefinitely by describing corrosion science and prevention literature, testing of physical and chemical properties, and many other subjects, but the works already listed will give an introduction to what is available and books on specific topics can be found reviewed in *WSCA*.

Directories

For paint manufacturers the most useful directory, despite its lack of an index, is the *Polymers Paint Colour Yearbook* (Read, 1986). For decorative paints there is the *Decorating Contractor Annual Directory* (Ridgway, 1986) and for industrial paints the *Finishing Handbook and Directory* (Bean, 1986). For UK resins there is the *British Resin Manufacturers' Association Index* (Cornelius, 1986). The US industry is dealt with in the *Paint Red Book* (Sweum, 1986), a 'comprehensive directory of formulators and suppliers to the coatings industry' and the National Paint and Coatings Association provides very detailed data on proprietary materials in their *Raw Materials Index*, with regular updates. Similar compilations for European raw materials, published every few years, are available on pigments and extenders (Luckert, 1980) and on resins and additives (Karsten, 1981). Directories for the paint industries of individual countries include the *Farbe und Lack Addressbuch* (published every 3 years), *France-Peinture* (every other year, CEPP) and *Repertorio dell' Industria Italiana dei Prodotti Vernicianti* (1986, ONEDIT).

For the adhesives industry two UK publications are the *BASA Yearbook* (annual, British Adhesives and Sealants Association) and the *Adhesives Directory* (Barraclough, 1986) while the *Adhesives Euro-Guide* (IAL, 1986) is both a commercial directory and a survey of the industry in different countries. The USA industry has its own directory (annual, Communications Channels Inc.)

Beyond these there is a range of directories covering the chemical industry, chemical engineering, building materials, brushes, concrete, packaging, sheet metals, laboratory equipment, market research, etc.

Conferences

At the American Chemical Society Meetings, every spring and autumn, many important topics first appear in public and in print. These are published first as *Polymer Preprints* (twice a year, American Chemical Society) and *Polymeric Materials Science and Engineering* (twice a year, American Chemical Society), both of which contain texts of papers presented at the Meetings, and also in abstract form in the *Book of Abstracts* (twice a year, American Chemical Society) which contains abstracts of all papers presented at the Meetings. Later the complete texts of papers from particular symposia within a Meeting may be published as individual books in the American Chemical Society Symposium Series.

An annual conference at which many important papers are given is the International Conference in Organic Coatings Science and Technology, held in Greece. The proceedings many not be published formally but all delegates receive preprints with complete texts of the papers, and the proceedings have been edited and republished, for a number of the earlier conferences, as *Organic Coatings Science and Technology* (Dekker). Another important annual event is the Waterborne and Higher Solids Coatings Symposium organized by the University of Southern Mississippi, of which the complete proceedings are published. Every other year the FATIPEC Congress is held in Europe, at which many useful papers are given, but the usual form of publication is preprint volumes provided to all delegates.

Many other conferences are held in various countries on specific subjects and the proceedings or preprints are important sources of information, while in some cases the individual papers are also published in journals. Abstracts of many conference papers are available in *WSCA* and in *RAPRA Abstracts*.

Patents

Patents are an important and often overlooked source of technical information. At present European patents are especially important, taken out by companies from all over the world to cover European countries as specified. Other important patents are those from the USA and from Japan, both representing active industries and major markets. Sources such as *WSCA* abstract thousands of patents every year, including also patents from other countries of importance. O'Neill (1979) provided a useful report on the importance of patents as a source of information for the paint industry, which is also valid for adhesives.

Standards

Within the United Kingdom, standards from the British Standards Institution are listed annually in the BSI Catalogue, and at irregular intervals BSI publish a Sectional List of standards on 'paints, varnishes, paint ingredients and colours for paints, etc.'

Unfortunately there is no Sectional List specifically for adhesives. Of universal importance are the International Standards from the International Standards Organisation, who issue a yearly catalogue with quarterly updates. Their *ISO Standards Handbook 24* (ISO, 1985) provides the complete texts of all ISO standards on paints and varnishes current at the time of publication. Individual countries provide catalogues of their own standards in various forms. The most convenient listing of newly published standards from all over the world is available from the BSI Library, as a monthly computer printout listing the titles of newly received standards, on whatever subject or combination of subjects the subscriber requests.

For the paint industry the most important and widely used are standards from the USA and from West Germany. USA standards relevant to the paint industry are summarized by the National Paint and Coatings Association in their *Guide to United States Government Paint Specifications*, containing both Federal and military standards, and updated at intervals. Of almost equal importance are the ASTM standards, published annually by the American Society for Testing and Materials in their *Annual Book of ASTM Standards*. Another USA organisation whose standards are used in many different countries is the Steel Structures Painting Council whose *Systems and Specifications*, republished at intervals, is the second volume of their *Steel Structures Painting Manual*. The German DIN standards are listed in *DIN-Katalog für Technische Regeln* (annual, Beuth Verlag) which also includes the titles of guidelines, standards, etc. issued by other official German organisations. A separate listing of DIN standards available in English translation is published periodically, as are collections of DIN standards, with the complete German text, on particular subjects, e.g. *Taschenbuch* 117 (1985) and 201 (1985) on 'raw materials for paints and coatings . . .'.

Two individual standards which are widely used are the European Scale of Degree of Rusting (from C.E.P.E.) and the Swedish Standard SIS 05 59 00 (1967) *Pictorial surface preparation standards for painting steel surfaces*.

Other standards which may be used include those from the European Coil Coating Association, including test methods, the UK Defence standards, and European standards.

Colour systems and standards

A particular problem for paints, possibly the first property associated with them, is that of colour. Unfortunately, the various colour standards are just collections of coloured cards, some of which are more systematic than others. BS 5252: 1976 *Framework for colour co-ordination for building purposes* contains 237 colours of which 100 also appear in BS 4800: 1981 *Specification for paint colours for building purposes* while BS 381C: 1980 *Specification for colours for identification, coding and special purposes* contains 107 colours, some but not all of which appear in BS 5252 or in BS 4800. The German RAL system is also widely used and contains several hundred colours, which may or may not appear in other standards. The Munsell colour system contains a series of about 1200 coloured chips, grouped according to hue, value and chroma with equal colour intervals, so that any colour can be given an approximate Munsell reference. Many colour standards quote Munsell references for their particular colours. Colour measurement values may also be quoted (tristimulus values, CIELab values, Hunterlab values, chromaticity co-ordinates, etc.) but these values depend on the instruments used to measure them.

Commercial information

Much company information, statistics, market information and business news is now available online, as described by Jotischky (1986). The sources for this range from directories to many journals, newspapers and company reports, some of which are abstracted in *WSCA* and in *RAPRA Abstracts*. A dated but still useful report on sources of statistics for the paint industry is that by Jotischky (1979) who has also produced a survey of statistics for the paint industries of many individual countries. Every other year Information Research Ltd. publish their *Profile of the United Kingdom paint industry* and *Profile of the European paint industry*, both of which collect together a large quantity of statistical data and information on market shares, industrial structure, trends, etc. For the US industry the equivalent publication is the *Rauch Guide* (Rauch Associates, 1985).

References

American Chemical Society. (1975). *Applied polymer science*. eds. Craver, J.K. and Tess, R.W. (American Chemical Society).

American Chemical Society. (2nd edn, 1985). *Applied polymer science* eds. Tess, R.W. and Poehlein, G.W. (American Chemical Society).

Association Quebécoise des Industies de la Peinture. (1977). *Vocabulaire Français/Anglais des peintures et revêtements; English/French paints and coatings vocabulary* (Association Quebécoise).

Bean, J.E. (1988). *Finishing handbook and directory 1988* (Sawell Publications Ltd).

Brewis, D.M. and Briggs, D. (1985). *Industrial adhesion problems* (Orbital Press).

British Adhesives and Sealants Association. (1988). *Yearbook and Directory* (Fuel and Metallurgical Journals).
British Maritime Technology Ltd. (1986). *Recommended practice for the protection and painting of ships* (BMT).
Chicago Society for Coatings Technology. (1980). *An infrared spectroscopy atlas for the coatings industry* (Federation of Societies for Coatings Technology).
Clausen, H. (1980). *Worterbuch der lacktechnologie* (Edition Lack und Chemie).
Communication Channels Inc. (1987). '1987 Adhesives Age directory'. *Adhesives Age*, **30**, (6), 39–214.
Cornelius, D. (1986). *Resin Index 1986* (British Resin Manufacturers' Association).
Créations Editions et Productions Publicitaires. (1988). *France-Peinture 1987–1988* (CEPP).
Dettner, H.W. (1971). *Elsevier's dictionary of metal finishing and corrosion in five languages* (Elsevier).
Editrice ONEDIT. (1988). *Repertorio dell'industria Italiana dei prodotti vernicianti* (Onedit).
Ernst, R. (1974). *Dictionary of industrial technics* (Oscar Brandstetter Verlag).
Gross, H. (1986). *Dictionary of corrosion and corrosion control* (Elsevier).
Hess, M. (1979). *Hess's paint film defects: their cause and cure*. ed. and rev. Hamburg, H.R. and Morgans, W.M. (Chapman and Hall).
Holman, R. (1984). *UV and EB curing formulation for printing inks, coatings and paints* (SITA Technology).
Hughes, J.F. (1984). *Electrostatic powder coating* (Research Studies Press).
Hummel, D.O. and Scholl, F.K. (1978–1984). *Atlas of polymer and plastics analysis* (3 vols). (Carl Hanser Verlag/Verlag Chemie).
Hurst, A.E. and Goodier, J.H. (9th edn, 1980). *Painting and decorating* (Charles Griffin).
IAL Consultants Ltd. (1986). *Adhesives Euro-Guide* (IAL Consultants).
Information Research Ltd. (1988). *Profile of the European paint industry* (Information Research Ltd).
Information Research Ltd. (1987). *Profile of the UK paint industry* (Information Research Ltd).
Jotischky, H. (1979). *Statistical sources for the paint industry* (Paint R.A.).
Jotischky, H. (1982). *The world paint industries: a review 1980–1981* (Paint R.A.).
Jotischky, H. (1986). *Business information on-line for the coatings industry* (Paint R.A.).
Karsten, E. (8th edn, 1987). *Lackrohstofftabellen*. ed. Luckert, O. (Curt R. Vincentz Verlag).
Lee, L.H. (1984). *Adhesive chemistry developments and trends* (Plenum Press).
Luckert, O. (1988). *Pigment und Fullstoff Tabellen* (M. und O. Luckert).
Mittal, K.L. (1983). *Adhesion aspects of polymeric coatings* (Plenum Press).
Morgans, W.M. (1982 and 1984). *Outlines of paint technology* Volume 1. Materials; Volume 2. Finished products. (Charles Griffin).
Munger, C.G. (1984). *Corrosion prevention by protective coatings* (National Association of Corrosion Engineers).
National Paint and Coatings Association. *Guide to US government paint specifications*. comp. Marsich, E.E. (N.P.C.A.)
National Paint and Coatings Association. *Raw materials index*. comp. Blankertz, E.E. (N.P.C.A.).
Oil and Colour Chemists' Association. (1976). *Introduction to paint technology* (OCCA).
Oil and Colour Chemists' Association Australia. (1983 and 1984). *Surface coatings*. Volume 1 Raw materials and their usage; Volume 2 Paints and their application (OCCAA).

O'Neill, L.A. (1979). *Patents, patent information and the paint industry* (Paint R.A.).
O'Neill, L.A. (1981). *Health and safety, environmental pollution and the paint industry* (Paint R.A.).
O'Neill, L.A. (1986). *Health and safety, environmental pollution and the paint industry: supplement to the second edition of 1981* (Paint R.A.).
Parfitt, G.D. (1973). *Dispersion of powders in liquids with special reference to pigments* (Applied Science).
Patton, T.C. (1973). *Pigment handbook.* Volume 1 Properties and economics; Volume 2 Applications and markets; Volume 3 Characterisation and physical relationships (Wiley).
Patton, T.C. (1979). *Paint flow and pigment dispersion: a rheological approach to coating and ink technology* (Wiley-Interscience).
Rauch Associates Inc. (1987). *Rauch guide to the US paint industry* (Rauch Associates).
Raaff, J.J. (1986). *Index vocabulorum quadrilinguis* (Van Goor Zonen).
Read, R. (1989). *Polymers Paint Colour Yearbook 1989* (Fuel and Metallurgical Journals Ltd).
Ridgway, L.V. (1988). *Decorating contractor annual directory 1988* (Ridgway Press).
Roffey, C.G. (1982). *Photopolymerisation of surface coatings* (Wiley).
Satas, D. (1986). *Plastics finishing and decoration* (Van Nostrand Reinhold).
Santholzer, R.W. (1969). *Five language dictionary of surface coatings, plating products, finishing, corrosion, plastics and rubber* (Pergamon Press).
Shields, J. (1984). *Adhesives handbook* (Butterworths).
Skeist, I. (1977). *Handbook of adhesives* (Van Nostrand Reinhold).
Society of Dyers and Colourists. (1982). *Colour Index: pigments and solvent dyes* (S.D.C.).
Solomon, D.H. and Hawthorne, D.G. (1983). *Chemistry of pigments and fillers* (Wiley-Interscience).
Steel Structures Painting Council. (1982 and 1987). *Steel structures painting manual.* Volume 1 Good painting practice; Volume 2 Systems and specifications (SSPC).
Sweum, A. (1988). *Paint red book* (Communication Channels Inc.).
Tekniska Nomenklaturcentralens. (1988). *Farg och lackteknisk ordlista* (AB Svensk Byggtjänst).
Turret-Wheatland Ltd. (1986). *Adhesives directory* (Turret-Wheatland).
Turner, G.P.A. (1988). *Introduction to paint chemistry and principles of paint technology* (Chapman and Hall).
Curt R. Vincentz Verlag. (1987). *Farbe und Lack Addressbuch* (Curt R. Vincentz Verlag).
Wake, W.C. (1982). *Adhesion and the formulation of adhesives* (Applied Science).
Weismantel, G.E. (1981). *Paint handbook* (McGraw-Hill).
Wyszecki, G. and Stiles, W.S. (1982). *Colour science — concepts and methods, quantitative data and formulae* (Wiley).

CHAPTER SIXTEEN (I)

New developments in polymers. Electronics and electrolytes

R.A. PETHRICK

Thirty years ago polymers were used almost exclusively in the electronics industry as insulators and dielectrics.[1] Dense non-polar materials such as polyethylene, polytetrafluorethylene and polystyrene can be used for the construction of co-axial cables, supports for wires and fed-through connections. Polymers such as polycarbonate, cellulose acetate and polyethyleneterephthalate, contain dipoles and can be used in the production of dielectric spaces for the construction of capacitors. Polyvinyl chloride, suitably plasticized is used as an outer insulating sheath for metallic conductors, however, it does suffer from the problem that if it becomes overheated it can give off toxic fumes.

About thirty years ago, it was recognized that conjugated organic structures should be capable of electronic conductivity, however, it is only in the last ten years that high values of electrical conductivity have been achieved in practice. The breakthrough in this area of research came firstly with the discovery of a relatively simple route to the synthesis of good quality films of polyacetylene using a Zeigler-Natta type of catalyst and secondly, the observation that the conductivity of these polymers can be increased by several orders of magnitude by the introduction of acceptor/donor dopant molecules.[2]

The polymeric conductor is essentially very different from a conventional inorganic semi-conductor in that they are intrinsically anisotropic and possess a quasi-one dimensional structure. The electron mobility within the conjugated chain is very high and the bulk conductivity is controlled by the inter-chain hopping processes. A wide range of polymeric structures have been shown to be able to support electronic conductivity. Polymers still have a very important role to play as dielectrics, insulators and are widely used as encapsulantsfor microelectronic components. In addition, they have played a key role in the lithographic steps upon which the construction of modern electronic components are based, as well as forming an important part in the construction of many sensor circuits. The following sections will attempt to survey each of these topics, however the reader is referred to the specialist literature collected at the end of this chapter for a detailed discussion on these topics.

Intrinsically conducting polymers

The simplest and most widely studied conjugated polymer is polyacetylene.[3, 4] It occurs in two thermally interconvertible forms, a cis and a trans form. The trans is the more stable structure and exhibits the highest electrical conductivity. Doping of thermally annealed polyacetylene films can lead to the generation of semi-conducting or metallic properties dependent upon the concentration and type of dopant used. Doping is usually achieved using either gaseous diffusion or electrothermal methods. The nature of the dopant governs whether p or n type material is produced: $((D^+)_y(CH))_x$ and $((A^-)_y\ (CH))_x$ where A is an acceptor and D is a donor. Semiconducting properties are achieved at approximately 1% dopant levels, higher dopant concentrations leading to metallic conductivity.

In an 'idealised' polyacetylene structure, all the carbon bonds would be identical in length. However, as a consequence of Peirels instability pairs of CH groups move towards each other forming shorter (double) and longer (single) bonds. Although the terms, single and double are used, the electron is not completely localised and the structure has features which resemble single and double bond characteristics. Conduction in an undoped polymer arising via the generation of random conformational defects which create localised non-linear topologically excited, soliton states. Raising the temperature increases the probability of such states occurring and leads to the generation of polaron states, through interaction of a neutral soliton and a charged 'anti' soliton state. The presence of a polaron generates two states in the band gap (a bonding and a non-bonding state) which are spaced equally either side of the Fermi level. The action of doping increases the number of polarons and these in turn can interact leading to the generation of 'free' solitons. The electron spin states of these various species are different and can be studied using electron spin resonance (ESR) and infrared spectroscopy.[5]

Polyacetylene has been the subject of a number of reviews and its physical, chemical and electronic properties are fairly well understood. [3,4,6–11] Despite the apparent attractiveness of organic semi-conductors and metals, commercial exploitation is likely to be limited by the fact that the localised electronic structures which give the polymer its electrical conductivity is also a potential source of instability. Exposure of a pristine film of polyacetylene to oxygen will initially lead to an increase in its conductivity, however, prolonged exposure allows chemistry to occur with a loss of conductivity due to the formation of addition structures and a decrease in the extent of a conjugation in the polymer.

A variety of applications has been proposed for organic conductors, including its use as a high power, high density, rechargeable storage battery. Doping of polyacetylene with lithium perchlorate in propylene carbonate generates a battery with an open circuit voltage of 3.7 V and a current of 0.01 A cm^{-2}. There has been considerable discussion of the relative merits of this system relative to more conventional lithium intercalation electrode systems and it is not clear at present time whether such a system is likely to be implemented commercially. This topic has been the subject of a recent detailed review by MacDiarmid[4] and it is clear that despite the intrinsic instability of the system it is possible to obtain some interesting properties. However, it does not appear that polyacetylene is unlikely to be a major commercial material despite the intense interest evident in the area of research between 1980 and 1985.

Other conducting polymer systems

The problem of stability which has limited the technological interest in polyacetylene can be improved by making the conjugation part of a series of conjugated ring systems, Figure 16(i).1. There are two basic forms of molecule possible,[12] when the R unit is a single heteroatom such as in the case of polypyrrole, or alternatively is part of an extended structure which may not contain heteroatoms, as in the case of polyazulene or polyparaphenylene. Maintaining the 'single-double' bond alternation whilst extending the separation between ring structures has also been attempted, a typical molecule being paraphenylene vinylene. The range of molecules capable of exhibiting electronic conductivity has been substantially extended and whilst they are unlikely to be applied either as semi-conductors or metals they are likely to be used as sensor elements in electronic devices.

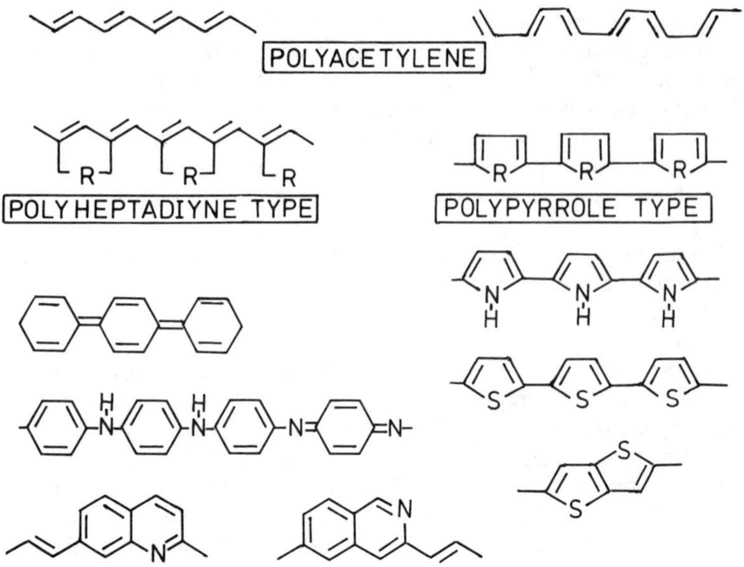

POLYACETYLENE

POLYHEPTADIYNE TYPE

POLYPYRROLE TYPE

Figure 16(i).1

i) Polyparaphenylene

Unlike trans polyacetylene, the two resonance forms of this polymer (benzenoid and quinoid) are not energetically equivalent with the result that the ground state is not degenerate and therefore there is not formation of a mid-gap state. However, as in the previous case, thermal excitation can lead to the formation of conduction. The effects of doping using Li and Na atoms has been explored and gaps of 0.64 and 0.93 eV respectively have been observed. ICI have found a biological route to the synthesis of this polymer and indicate that such systems can be generated via other routes than the analogue of polyacetylene.

ii) Polydiacetylene

The properties of diacetylene and the related polydiacetylene have been studied for many years.[12] The monomer can be grown in the form of highly perfect single crystals which undergo topochemically controlled reaction to form the polymer. Whilst the intrinsic electron mobility within the chains is high, the overall electrical conductivity lies between 10^{-6} to 10^{-12} mhos cm^{-1} depending upon temperature. Numerous attempts have been made to produce metallic conductivity in the polymer form, however, doping of these systems appears to be intrinsically difficult. Soluble polydiacetylene compounds have been synthesized and also used to grow Langmuir Blodgett films, however, they are no more attractive than polyacetylene or polyparaphenylene as organic conductors.

iii) Polypyrrole

Unlike the systems presented above, polypyrrole has the attraction of containing a heteroatom which imparts stability and also the possibility of doping with relatively inert species. Although the heteroatom does not take any significant contribution to the conduction band, they do modify the valence band, leading to an asymmetrical effect upon any bonding/non-bonding states which may occur in the band gap. Polypyrrole has been observed to have a band gap of about 3.2 eV and shows ESR behaviour similar to that of polyparaphenylene.[12] Highly conducting films have been obtained from polypyrrole which electronically are comparable to polyacetylenes, but have greater oxidative stability.

One attraction of polypyrrole is that it can be synthesized electrochemically. The polymer forms at the anode and incorporates the anion in a ratio normally between 1:3 and 1:4 (anion:monomer). A substantial number of media have been used as the electrolyte and include benzonitrile, propylene carbonate and water-acetonitrile. A wide range of salts have been examined as electrolytes and include $AgC1O_4$, $LiBF_4$ and t-butylammonium toluene-*p*-sulphonate as well as a series of n-alkyl sulphonates.[14] The latter have a dramatic effect on the structure and electrical properties of the resultant material. As the alkyl chain length is increased so the electrical conductivity decreases indicating a reduction in the probability of electron chain to chain transfer. It has also been observed that the conductivity of polypyrrole synthesized electrochemically depends upon the doping level and also on the voltage used.[15] Maximum conductivity is achieved when the dopant level approaches the 1:3 level and also when the potential is close to the redox values for the film. It is also observed that the number of protonated to non-protonated rings is very dependent upon growth conditions and is directly correlated with the conduction level. Treatment of pristine films with sodium hydroxide leads to an approximately five-fold decrease in the conductivity, which can be restored by treatment with hydrochloric acid.[16]

Polypyrrole films have also been used extensively by electrochemists wishing to produce films which are capable of being either a source of electrons or mobile anions. A typical film system will be composed of a system capable of undergoing a redox process and this will be in contact with a metallic electrode assembly. For example, during the oxidation of a film of poly($Os(bpy)_2(vpy)_2)(ClO_4)_2$), additional anions will migrate into the film from a contacting electrolyte solution; poly($Os(bpy)_2(vpy)_2)(ClO_4)_2$) + ClO_4 → poly($Os(bpy)_2(vpy)_2$)-($ClO_4)_3$) + e^-. The state of the polymer film can be assessed by either cyclic voltametry or by measurement of the current voltage characteristics of the films. In this situation the conducting polymer films are not carrying a large current and hence have a relatively long and stable life. A considerable number of systems have been devised and a number of detailed reviews have been published on this topic.[17,18] A more direct use of polypyrrole has been proposed, in which the conducting polymer is used to sense the production of protons by an enzyme which is in contact with the solution being analysed. The conductivity of the polymer film is therefore directly related to the concentration of a particular species in solution and we have the principle of a useful biosensors system.

Polymer electrolytes

One of the few examples of the use of a polymer as a current device is poly-2-vinylpyridine which with lithium iodide forms the main constituent of a heart pacemaker power source.[19] One of the main problems encountered when designing a polymer electrolyte is the volume change which occurs during charging and discharge. Solutions to this problem have been sought by using i) soft, deformable solid components; ii) internal cell configurations that adapt to volume effects which do not lead to cell failure; iii) good adhesion; and iv) use of spring loading. Despite these problems the above system works sufficiently well to be used in most pacemakers. The life of this battery is approximately five years and then a minor surgical operation is required to replace the power source.

Recently much interest has been shown in other polymer systems and a wide range of systems have been investigated; however, the most attractive at the present time is polyethylene oxide.[20] This polymer when heated to a temperature of approximately 110°C has sufficiently high cation mobility to allow its use in both primary and secondary battery applications. The current research activity is directed towards the production of a room temperature high mobility cation conduction medium. The conductivity of polyethylene oxide can be improved by lowering its crystallinity either by the addition of plasticizing agents or by crystallizing it into an amorphous form. If a suitable material can be achieved it should be possible to construct an all-solid-state battery rather on the lines of that proposed by Harwell in their battery project.[21] The advances which have been made in the last five years indicate that this objective is probably achievable and this may be one of the first of the major uses of polymers in an active sense in electronic devices.

Polymers used in micro-lithography

One of the major changes which has occurred in the last ten years in the design and construction of electronic circuits has involved the reduction in size of the components which form the circuit elements from approximately 100 to 1 micron feature size.[22–24] This reduction has been achieved by the use of lithography based on polymeric resist materials. The resist must be able to respond to light or electron beams to produce a pattern which is capable of being transferred by etching into the substrate or protects the substrate from damage during a metal deposition or ion implantation step. The most widely used resists are based on novolac resins which are rendered insoluble to the alkali by dispersing in the matrix an azide dye. On exposure of the resist to light, the azide is transformed into an alkali-soluble carboxylic acid which allows pattern generation to be achieved by solvent development. Dyes can also undergo decomposition to give radicals can induce crosslinking of modified polybutadiene polymers and this forms the basis of another widely used resist material. Exposure of resists to electron beams can induce either crosslinking or degradation. In the case of polymethyl methacrylate and related materials the exposure process leads to a reduction in the molecular weight of the polymer and there is an increase in the solubility of the exposed area. In the case of halogenated styrene based polymers, exposure of the resist to the electron beam induces crosslinking and this renders the exposed area less soluble than the surrounding matrix.

A number of articles have been published on the chemistry of micro-lithography[22–24] and with the industrial demand for materials with plasma etch resistance and ever decreasing feature size we can anticipate that this will be an active area of research for a number of years to come. A selection of review articles are presented in the bibliography at the end of this chapter.[22–24]

Polymeric insulators

In the context of microcircuit fabrication considerable interest has been paid into the characteristics of polyimide films.[25,26] These have excellent insulation characteristics and are able to stand up to the aggressive atmosphere used in microelectronic circuit production. The films are often produced by first depositing the precursor polymer which on heating turns into the high temperature rigid form. Much effort has gone into the formulation of suitable precursors and with the advances in thin film deposition we are likely to see further interest in this area of study.

Epoxy and silicon resins have been used for many years as encapsulants for microcircuits. Research still continues in this area to overcome the problems of thermal expansion, moisture transport inhibition and the electrolytic processes which occur in such systems. Here the polymers appear to play a passive role, however, it is clear that suppression of moisture infusion into a potted resin requires the use of active ingredients in the matrix.[25]

Polymers as electronically active elements

The main use of active polymeric elements in the electronics industry is as screens for high frequency electrical signals and static electrical charges. These screens are often formed by the incorporation of either carbon black or metal particles or layers into a polymer resin. The conducting paths of the carbon black provide the matrix with the capability to absorb radio signals or to provide a conducting path to ground. Large amounts of polymeric materials are used in this area of circuit packaging. Circuits have also been devised in which carbon black loaded crosslinked polymer materials form a conducting element between two metal feed wires. The passage of a current through the matrix causes heating and this results in an expansion of the matrix and a reduction of the heating output. It is therefore possible by suitable selection of the polymer matrix, its crosslink density and the level of carbon black incorporation to produce a circuit element which self-regulates the power output of the heating circuit so as to achieve a fairly constant external temperature. Commercial devices using this configuration have been in use for a number of years.

Another wide use of polymer composites is in reprographics.[27,28] Photocopying is achieved by the development of an image in a photoconducting photoreceptor by cascading across its surface small particles of polymer pigmented with carbon black.[29] The design of the pigmented articles is by no means a trivial task and the electrical properties of these materials have been studied intensely for many years. It is now clear that the charging ability of the pigment particles is connected with the extent to which the carbon black particles form aggregated structure. Contact between the toner particles and a metal, can, through contact charging, generate a charge and this can be adjusted to be equal and opposite to that on the photoreceptor and thus allows the image to be visualised. The image is then transferred to paper to produce a copy of the original by the application of a potential which is larger than that associated with the image on the photoreceptor. The final stage in the process is heating the polymer beads to a point where they fuse with the paper to give the fixed image. The process has become so much part of our lives that we scarcely appreciate that it relies on a very subtle balance of electronic properties of a polymer composite.

Piezoelectric polymer films

Over the last ten years research has been carried out into the piezoelectric characteristics of aligned poled films of polyvinylidene fluoride.[31] This polymer crystallizes into an asymmetric structure which on application of an electric field will lead to a mechanical displacement. Polyvinylidene fluoride unfortunately is not as efficient as many of the traditional inorganic materials and has therefore not made a significant commercial impact, however, it is being used in situations where large area sensors are required in contact with aqueous media. Research continues in this area and it is to be hoped that a more efficient material may emerge in the future allowing polymers to be more effectively used as active elements in electronic applications.

Whilst it is rather unlikely that we shall see conducting polymer replacing silicon or metals, there is a greater probability that we shall see them used either as part of composite electrodes in light-weight battery assemblies or as part of a biosensor device.[8–11] Polymers are already playing an active role as polymeric electrolytes in certain battery systems and are likely to be the material used in the construction of solid-state battery systems.

The above short review has hopefully indicated the way in which polymers are making an impact in a number of areas because of their electrical characteristics and this list will probably grow and broaden in the future.

References

1. Skirakaura, H., Louis, E.J., MacDiarmid, A.G., Chiang, C.K. and Heeger, A.J. (1977). *J. Chem. Soc. Chem. Commun.*, 578.
2. Nigrey, P.J., MacDiarmid, A.G. and Heeger, A.J. (1979). *J. Chem. Soc. Chem. Commun.*, 594.
3. Simon, J. and Andre, J.J. (1985). *Molecular semi-conductors*, (Springer-Verlag).
4. MacDiarmid, A.G. and Maxfield, M.R. (1987). *Electrochemical science and technology of polymers-1* (Elsevier Applied Science Publishers), 67.
5. Shacklette, L.W. and Chance, R.R. (1979). *Synth. Mat.* **1**, 307.
6. Kuzman, H., Mehring, M., and Roth, S. (1985). *Electronic properties of polymers and related compounds, solid state sciences 63* (Springer-Verlag).
7. Pethrick, R.A. and Mahboubian-Jones, M.G.B. (1985). *Polymer Yearbook 2*, Pethrick, R.A. (ed), 201.
8. MacDiarmid, A.G. and Heeger, A. J. (1980). *Organic Metals and Semi Conductors, Synthetic Metals*, **1**, 101.
9. Nigrey, P.J., MacInnes, D. Nairns, D.P., MacDiarmid, A.G. and Heeger, A.J. (1981). *Conductive Polymer*, R.B. Seymour (ed) (Plenum Press, New York).
10. Street, G.B. and Clarke, T.C. (1981). Conducting polymers: A review of recent work, *IBM, J. Res. Dev.*, **25**, 51.
11. Boughman, R.H., Brodas, J.L., Chance, R.R., Esibaumer, R.L. and Shacklette, L.W. (1982). *Chem. Rev.* **82**.
12. Bedas, J.L. (1984). *J. Mol. Structure (Theochem)*, **107**, 169.
13. Bloor, D. (1981). *Photon, electron and ion probes of polymer structure and properties*, (ACS Symposium Series), **162**, 65.
14. Wernet, W. Monkenbusch, M. and Wegner, G. (1984). *Macromol. Chem.* **5** (3), 157.
15. Al-Arrayed, F.M., Bonham, H. McLeod, G., Mahboubian-Jones, M.G.B. and Pethrick, R.A. (1986). *Materials Forum*, **9**, No.4, 209.
16. Pethrick, R.A., Jeffrey, K., McLeod, G. and Affrossman, S., unpublished data.
17. Murray, R. and Chidsey, C.E.D. (1986). *Science*, **231**, 25.
18. Hillman, A.R. (1987). *Electrochemical science and technology of polymers-1*, Linford, R. (ed) (Elsevier Applied Science Publishers), 103.
19. Owens, B. (1985). *Solid state batteries*, (NATO ASI No 101, Series E), 389.
20. Armand, M.B. 1980). *Lithium non-aqueous batteries electrochemistry*, Pennington, H. (ed) (The Electrochemistry Society), **80**,(7), 261.
21. Sequeira, C.A.C. and Hooper, A. (1985). *Solid state batteries* (NATO ASI No 101 Series E), 402.
22. Thompson, L.F., Willson, C.G. and Fechet, J.M.J. (1984). *Materials of microlithography* (ACS Symposium Series 219).
23. Thompson, L.F. Willson, C.G. and Bowden, M.J. (1983). *Introduction of microlithography* (ACS Symposium Series 219).
24. Pethrick, R.A. (1986). *Progress in Rubber and Plastics Technology*, **2**, 1–9.
25. Feit, E.D. and Wilkins, C. (1982). *Polymer materials for electronic applications* (ACS Symposium Series 184).
26. Giess, E.A., Tee, K.N. and Ullmann, D.R. (1985). *Electronic packaging materials science* (Materials Research Society, Vol 40).
27. Davis, D.K. (1982). *Electrical properties of polymers*, Seanor, D.A. (ed) (Academic Press), 285.
28. Hair, M. and Croucher, M.D. 1982). *Colloids and surfaces in reprographic technology* (ACS Symposium Series 200).

29. Duke, C.B. (1982). *Photon, electron and ion probes of polymer structure and properties* (ACS Symposium Series 200).
30. Daly, J.H., Hayward, D., and Pethrick, R.A. (1986) *J. Phys. D*, **19**, 885.
31. Kawai, H. (1969). *Jap. J. Appl. Phys.*, **8**.

CHAPTER SIXTEEN (II)

New developments in polymers. High performance plastics

R.J. CRAWFORD

1.0 Introduction

Plastics are nowadays widely accepted as materials which may be used in demanding, load-bearing applications. This is a considerable improvement on the situation which existed even a relatively small number of years ago, when plastics were regarded as only being suitable for simple domestic articles. However, engineers and chemists are not content to settle for their achievements so far and there are continuing efforts to develop plastics with even better performance characteristics. Polymers are in fact a unique class of materials because their almost infinite versatility in regard to synthesis means that they can be tailor-made for almost any application. Economic factors place a restriction on the number of polymers which are commercially available but if there is sufficient demand then new plastics can be developed not only through physical compounding but also through adjustment of the molecular architecture.

2.0 Development of high performance plastics

The first level at which the properties of polymers can be controlled is through molecular design. Polymers consist of long chain-like molecules and the properties of the plastic are controlled by the average length of the chains, the stiffness of the individual chains, the strength of the forces between the chains and the regularity with which the chains can pack together. The strongest intermolecular forces arise when the chains are designed to link together chemically. This process of cross-linking is usually arranged to occur during the final stages of the shaping of the plastic article. The end-product is thus made from a material which exhibits the characteristics of being very stiff and resistant to high temperatures because the chemical cross-linking is not reversible by the application of heat. Plastics of this type are termed *thermosets*. Examples of these materials are phenolics (Bakelite), epoxies and some types of polyester. These were the first polymers to exhibit high performance specifications. Rubbers, both synthetic and natural, also develop cross-links during moulding and this enables them to withstand high temperature and exhibit elastic recovery after stress removal.

In contrast to this type of cross-linked structure, there is another important class of polymer in which the individual molecular chains are in close proximity but are not joined together. These materials are termed *thermoplastics* and since the molecular chain mobility increases with temperature, these materials will soften and become flexible when they are heated. Examples of thermoplastics are nylon, acrylic, and polystyrene.

In the thermoplastic range of materials, high performance characteristics can be achieved through control of the molecular morphology or through physical combination with other materials. In the case of polypropylene, for example, the molecular chain is intrinsically flexible but the material can be made quite stiff through crystallization i.e. orderly close packing of the molecular chains. However, further improvements in stiffness can only be achieved through combination with other stiffer and stronger materials, such as fibres. The use of fibres in polypropylene would put it into the category of *engineering thermoplastics* i.e. those materials such as nylon, acetal, polycarbonate, modified polyphenylene oxide, polysulphone, thermoplastic polyesters and polyphenylene sulphide which earn this title because they can withstand modest loads more or less indefinitely. However, to be considered as an *advanced polymer* it is necessary to have high load carrying capability in a demanding environment (usually high temperature). To satisfy these requirements it is necessary to look to the latest generation of polymers and composites.

One of the most recent developments in molecular design is a chain structure which is inherently more straight and rigid than the normal types of polymer chain. Aromatic groups incorporated into the chain serve this purpose because they act as stiffening elements. The molecular chains thus have a rod-like character and in some circumstances the rods can be aligned to act like fibres. In other cases the structure may be more random but the rod-like elements lock together to offer increased stiffness. Polyetheretherketone (PEEK) is an aromatic thermoplastic which has already earned itself a reputation as one of the new high performance plastics. In the unreinforced form it may be used continuously at 160°C (it melts at 334°C) and when glass or carbon fibres are added, the service temperature can be increased to more than 300°C. A more recent aromatic thermoplastic (polyetherketone, PEK) offers even better high temperature performance in that the unreinforced grades can be used continuously at 250°C. It is also worth noting that aramid fibres, which have a tensile strength comparable to steel and are used to reinforce plastics, have a structure which is based on the alignment of rod-like chain elements.

Although the characteristics of individual plastics can be modified by blending or alloying with other plastics this only tends to produce a better combination of properties in the final material rather than providing a major improvement in performance. In order to achieve the latter it is necessary to add fibres to the plastic. The fibres normally used are glass, aramid or carbon in either chopped or continuous form and they may be used in thermosets and thermoplastics. The final material produced in this way is called a *composite*. In some cases the term 'compound' is used if the fibres are very short and the term 'composite' is reserved for situations where the fibres are very long or continuous. Alternatively, any material reinforced with fibres (short or long) may be referred to as a composite and the term 'advanced composite' is then used to distinguish those materials with very high performance characteristics. The advanced composites are capable of competing favourably with metals in the most demanding application areas, for example, primary aircraft structures and aerospace components.

Although both thermoplastics and thermosets can benefit from fibre reinforcement, they have tended to develop in separate market sectors. This situation has arisen due to the fundamental differences in the nature of the two classes of material, both in terms of properties and processing characteristics.

Thermosets have tended to have their application areas limited by their inherent brittleness and maximum property enhancement was sought through the use of continuous fibres (in filament form or woven into mats). Their major advantage here is that their low viscosity prior to cross-linking enables excellent impregnation of the fibrous mat. The thermoplastics industry has, on the other hand, seen its advantage in fast production rates through the use of injection moulding, for example. However, this has restricted them to the use of short fibres with a consequent sacrifice in properties.

Nowadays, however, the two material sectors are moving closer together and are competing on one another's territory. The thermosetting industry is prepared to reduce fibre length in order to achieve faster production rates and thus open up new market areas. Similarly, the thermoplastics industry is seeking a niche in the aerospace market by including very long or continuous fibres in PEEK, for example, and shaping the material by slower methods such as compression moulding.

3.0 Sources of information on high performance plastics

The previous section has set the scene for the sort of developments which are currently taking place in the field of high performance plastics. Although these are new materials there is in fact a vast amount of information which is available and it would not be possible to review it all or even list it all in the limited space available here. However, the following section should assist the reader to make a start in getting familiar with the type of information which is available. This may be obtained mainly from learned societies, research associations, universities and polytechnics, books, trade literature, journals and specific research papers. For convenience the latter area has been subdivided into general papers and those in the automotive and aerospace industries since these tend to be main categories of users of high performance plastics.

Information on individual engineering thermoplastics may be obtained from the manufacturers. Table 16(ii).1 indicates the European suppliers of each of the materials.

Journals

Polymer Contents Fried, J.R. (ed) (Elsevier Applied Science, London) (monthly). This journal provides details of the contents of over 90 polymer related journals.

High Performance Plastics Juniper, R. (ed) (Elsevier Applied Science, London in association with RAPRA, Shawbury) (monthly).

Composites Science and Technology Harris, B. (ed) (Elsevier Applied Science, London) (monthly).

Composite Structures Marshall, I.H. (ed) (Elsevier Applied Science, London).

New Materials/Japan Shibuya, H. (ed) (Elsevier Applied Science, London) (monthly).

Table 16(ii).1 Key suppliers of engineering plastics in Europe

Supplier	Country	ABS	Acetal	Polyamide	Polycar-bonate	Polyester	PPO (mod)	Poly-sulphone	Polyamide-imide	Polyether-imide	Polyether-sulphone	Polyphenylene sulphide	PEEK
Amoco	Switzerland	—	—	—	—	—	—	✓	✓	—	—	—	—
Atochem	France	✓	—	✓	✓	✓	—	—	—	—	—	—	—
Akzo	Norway	—	—	✓	—	✓	—	—	—	—	—	—	—
BASF	Germany	✓	—	✓	—	✓	—	✓	—	—	✓	—	—
Bayer	Germany	✓	—	✓	✓	✓	—	—	—	—	—	✓	—
BIP	UK	—	—	✓	—	✓	—	—	—	—	—	—	—
Borg Warner	Norway	✓	—	—	—	—	—	—	—	—	—	—	—
Hoechst	UK	—	✓	✓	—	✓	—	—	—	—	—	✓	—
CW Huls	Germany	—	—	✓	—	✓	—	—	—	—	—	—	—
CdF Chimie	France	✓	—	—	—	—	—	—	—	—	—	—	—
Ciba Geigy	Switzerland	—	—	—	—	✓	—	—	—	—	—	—	—
DSM	Norway	✓	—	—	—	—	—	—	—	—	—	—	—
Du Pont	Switzerland	—	✓	✓	—	✓	—	—	—	—	—	—	—
DOW Chemical	Switzerland	✓	—	—	—	—	—	—	—	—	—	—	—
Eastman	Switzerland	—	—	—	—	✓	—	—	—	—	—	—	—
EMS Chemie	Switzerland	—	—	✓	—	—	—	—	—	—	—	—	—
Enichem	Italy	✓	—	✓	✓	✓	—	—	—	—	—	—	—
Gen. Electric	Norway	—	—	—	✓	✓	✓	—	—	✓	✓	—	—
Hoechst	Germany	—	✓	—	—	—	—	—	—	—	—	—	—
ICI	UK	—	—	✓	—	—	—	—	—	—	✓	—	✓
Monsanto	Belgium	✓	—	✓	—	—	—	—	—	—	—	—	—
Montepolimeri	Italy	—	—	✓	—	✓	—	—	—	—	—	—	—
Phillips Pet.	Belgium	—	—	—	—	—	—	—	—	—	—	✓	—
Rhone-Poulene	France	—	—	✓	—	✓	—	—	—	—	—	—	—

Composites Adams, M. and Herriot, J. (eds) (Butterworths, London) (quarterly).
Additives for Polymers (Elsevier Applied Science in association with Yarsley Technical Centre) (monthly).
Composites and Adhesives Newsletter (T/C Publications, Pasadena) (bimonthly).
Advanced Composites Engineering (Design Council, London) (quarterly).
Polymers/Ceramics/Composites Alert (Materials Information, a joint service of the ASM International and the Institute of Metals) (monthly).

Books

Balasubramanian, N. (ed) (1987). *Advanced composite materials index 1975–1984* (Technomic Publishing AG, Basel).
Benjamin, B.S. (1982). *Structural design with plastics*, (Van Nostrand, New York).
Bower, C.M. (ed) (1985). *Composite material glossary*, (T/C Publications, Pasadena).
Bunsell, A.R. and Kelly, A. (eds) (1985) *Composite materials: a directory of European Research* (Butterworths, London).
Ciferri, A. and Ward, I.M. (eds) (1979). *Ultra high modulus polymers*, (Elsevier Applied Science, London).
Clegg, D.W. and Collyer, A.A. (eds) (1986). *Mechanical properties of reinforced thermoplastics* (Elsevier Applied Science, London).
Crawford, R.J. (1985). *Plastics and rubber: engineering design and applications* (MEP Ltd., Bury St. Edmunds).
Fitzer, E. (ed) (1985). *Carbon fibres and their composites*, (Springer-Verlag, Munich).
Folkes, M.J. (1982). *Short fibre reinforced thermoplastics*, (Research Studies Press, London).
Gill, R.M. (1972). *Carbon fibres in composite materials*, (Iliffe, London).
Gotham, K.V. and Hough, M.C. (1984). *Durability of high temperature thermoplastics* (RAPRA, Shrewsbury).
Halpin, J.C. (1984). *Primer on composite materials: analysis* (Technomic Publishing AG, Basel).
Hancox, N.L. (ed) (1981). *Fibre composite hybrid materials* (Applied Science, London).
Herriot, J. (ed) (1987). *Composites evaluation: proceedings of the second international conference on the testing, evaluation and quality control of composites (TEQC 87)* (Butterworths, London).

Houldcroft, P.T. (ed) (1987). *Material data sources* (MEP Publications Ltd., London).
Hussein, R. (1986). *Composite panels and plates*, (Technomic Publishing AG, Basel).
Martuscelli, E., Marchetta, C. and Nicolais, L. (1987). *Future trends in polymer science and technology* (Technomic Publishing AG, Basel).
Mathews, F.L. (1987). *Joining fibre reinforced materials* (Elsevier Applied Science, London).
Meyer, R.W. (1986). *Handbook of pultrusion technology*, (Chapman and Hall, London).
Meyer, R.W. (1987). *Handbook of polyester moulding compounds* (Chapman and Hall, London).
Pritchard, G. (ed) (1986). *Developments in reinforced plastics* (Applied Science, London).
Richardson, M.O.W. (1977). *Polymer engineering composites* (Elsevier Applied Science, London).
Seymour, R.B. and Kirschenbaum, G.S. (eds) (1986). *High performance polymers* (Elsevier, Amsterdam).
Sheldon, R.P. (1982). *Composite polymeric materials* (Elsevier Applied Science, London)
Shool, G.D. (1986). *Reinforced plastics for commercial composites — source book* (Am.Soc. of Metals, Metals Park, Ohio).
Titow, W.V. and Lanham, B.J. (1975). *Reinforced thermoplastics* (Applied Science, London).
Tsai, S.W. and Hahn, H.T. (1980). *Introduction to composite materials* (Technomic Publishing AG, Basel).
Composite Market Reports Inc., *Advanced composites materials key personnel list* (T/C Publications, Pasadena).
Advancing technology in materials and processes (Technomic Publishing AG, Basel, 1985).
Directory of books on plastics (T/C Publications, Pasadena).
European centres of development on advanced composite materials (Metra Martech, London, 1987).
Fibre reinforced composites 1986 (1986). Proceedings of Institute of Mechanical Engineers Conference, 1986. (MEP Publications Ltd., London).
Worldwide carbon fibre directory (Pammac Directories Ltd., London).

Useful addresses

The British Plastics Federation, 5 Belgrave Square, London SW1X 8PH, UK.

The Plastics and Rubber Institute, 11 Hobart Place, London SW1W 0HL, UK.
Rubber and Plastics Research Association (RAPRA), Shawbury, Shrewsbury, Salop SY4 4NR, UK.
INSPEC (Advanced Materials Abstracts), The Institution of Electrical Engineers, Station House, Nightingale Road, Hitchin, Hertfordshire SG5 1RJ, UK.
The British Composites Society, c/o Mr F Mathews, Imperial College of Science and Technology, London SW7, UK.
University Microfilm International Doctoral Dissertations, White Swan House, Godstone, Surrey RH9 8LW, UK.

Research papers

(i) General

'Engineering and plastics', *Engineering*, (July/August. 1985).
Chalkley, R.M. (1981). 'Injection moulded gears', *CME*, June, p43.
'Plastics in ball bearings', *Ball Bearing Journal No 228*, (November, 1986).
Belbin, G.R. (1986). 'Engineering thermoplastics: looking towards the future', *Plastics and Rubber International*, December, p29.
Mitchell, P.D. (1980). 'Engineering plastics', *Engineering*, October, p1127.
Caesar, H.M. (1981). 'Engineering thermoplastics: an assessment', *Metallurgist and Materials Technologist*, May, p265.
Caesar, H.M. and Blumberg, H. (1986). 'Growth prospects for high strength compounds and composites', *Modern Plastics International*, October, p145.
'Speciality thermosets', *Modern Plastics International* (September, 1985), p78.
'Engineering thermoplastics', *Plastics and Rubber Weekly*, Special Report (February, 1987), p12.
Blinne, G. and Theysohn, R. (1986). 'Engineering thermoplastics, today and tomorrow', *Modern Plastics International*, October, p121.
Baer, E. (1986). 'Advanced Polymers', *Scientific American*, October, p157.
Chou, Tsu-Wei, McCullough, R. and Pipes, R.B. (1986). 'Composites', *Scientific American*, October, p167.
Molyneux, M. (1983). 'Technology for advanced composites', *Composites*, **14** (2), p87.

Dodds, R. (1983). 'Designing in reinforced thermosets', *Engineering*, September, p1.
Whelan, A. and Goff, J. (1986). 'Filled engineering thermoplastics', *British Plastics and Rubber*, February, p29.
Schmelzer, E. (1986). 'Automation for SMC parts', *Modern Plastics International*, November, p71.
Wood, A.S. (1986). 'The majors are taking over in advanced composites', *Modern Plastics International*, April, p40.
Vinson, J.R. (1986). 'Recent advances in technology for composites in the USA', *Materials and Design*, January/February, p6.
Batchelor, J. (1981). 'Use of fibre reinforced composites in modern railway vehicles', *Materials in Engineering*, **2**, June, p172.
Waterman, N.A., Trubshaw, R. and Pye, A.M. 'Filled thermoplastics' *Materials in Engineering Applications*, Part I, Vol I, December (1978) p74: Part II, Vol I, June (1979) p203.
Green, A.K. and Phillips, L.N. (1978). 'Non-aerospace applications for high performance materials', *Materials in Engineering Applications*, **I**, December, p59.
King, R.L. (1982). 'A production engineer's view of advance composite materials', *Materials and Design*, **3**, August, p515.

(ii) Automotive

'Plastics in passenger cars', SP-566, Society of Automotive Engineers, Warrendale, Pennsylvania (1984).
Trewin, E. (1981). 'Potential for carbon fibres in automotive applications', *Plastics and Rubber Processing and Applications*, **1**(2), p101.
Sternfield, A. (1986). 'Top-of-the-line engineering plastics move into high series applications', *Modern Plastics International*, June, p40.
Dreger, D.R. (1984). 'Automakers continue to say yes to plastics', *Machine Design*, October, p56.
Maxwell, J. (1983). 'The drive for plastics', *Plastics and Rubber International*, April, p45.
'Plastics at the heart of the engine', *Engineering*, (April, 1986), p313.
Waterman, N.A. (1984). 'Plastics replace metals', *Engineering*, May, p370.
Mair, H.J. (1984). 'Plastics in automotive manufacture, *Industrial and Production Engineering*, March, p104.
'Opportunities for plastics in 1990 model cars', *European Plastics News*, (May, 1986), p12.
McGeekin, P. (1942). 'Composites in transportation' *Materials in Engineering*, **3**, April, p378.

Charlesworth, D. (1981). 'Potential use for plastics in automobiles', *Materials in Engineering*, **2** March, p149.
Ruegg, C. (1983). 'Carbon and aramid reinforced plastics in the manufacture of automotive propeller shafts', *Materials and Design*, **4**, August/September, p813 and October/November, p870.

(iii) Aerospace

Noyes, J.V. (1983). 'Composites in the construction of the Lear Fan 2100 aircraft', *Composites*, **14**(2), p129.
Riley, B.L. (1986). 'AV-8B/GR Mk 5 Airframe composite applications', *Proc. I. Mech. E.*, **200** No(50).
'Applications for composites', SRI Int. Business Intelligence Program, Report No 700 (1984).
'Composites take off', Engineering, (September, 1982), p611.
Thomas, D.K. (1983). 'The uses of rubber and composites in aerospace', *Plastics and Rubber International*, April, p53.
Oliver, D. (1983). 'France's reinforced plastics helicopters', *IRPI*, November/December, p10.
'Certification of first CFRP propeller', *IRPI*, (May/June, 1984), p14.
Russell, J.G. (1985). 'Aircraft propulsion comes the full circle', *CME*, April, p37.
Emery, R.K. and Graham, R. (1985). 'Fibre reinforced plastics for space structures', *CME*, April, p47.
Beercheck, R.C. (1984). 'Superchoppers are on the way', *Machine Design*, September, p84.

PART III

CHAPTER SEVENTEEN

Europe (excluding the UK, Scandinavia and the CMEA countries

P. KÄFER

This chapter gives a survey of European information sources in polymers excluding UK, Scandinavian and Eastern Bloc sources. Such information sources include journals, books, abstracting services, conferences and data bases. This chapter is mainly concerned with specific polymer information sources, but general information sources which mention polymers are also included.

Books and reviews

It is of course impossible to give a complete survey of books and reviews in the polymer field in Europe in this chapter. But there are some reviews of special interest which should be mentioned.

The *Informationsführer Kunststoffe* (P. Eyerer, VDI-Verlag GmbH, Düsseldorf) is a guide to literature search and provides information on monographs, journals, abstracting services, organisations, standards etc. in the polymer field worldwide. The *Literaturführer Kunststoffe 1945–1980* (P. Eyerer) covers monographs published throughout the world, and university treatises published in West Germany and the USA. It is an updated and greatly expanded section of the *Informationsführer Kunststoffe*.

An introduction to polymer chemistry is given by H.G. Elias in *Makromoleküle* and by H. Baker and F. Lohse in *Einführung in die Makromolekulare Chemie* (Hüthig Verlag GmbH, Heidelberg).

A detailed survey of polymer chemistry can also be found in Houben-Weyl *Methoden der organischen Chemie* (Thieme Verlag, Stuttgart). A completely new edition of the polymer volumes of 'Houben-Weyl' has been published in 1987.

The *Kunststoff-Handbuch* (G. Becker, D. Braun, Carl Hanser Verlag, München) contains a detailed description of the different polymer types (polyolefins, polyurethanes, polyesters, etc.). The *Kunststoff-Taschenbuch* (H. Saechtling, Carl Hanser Verlag), which is also edited in English, covers the whole field of plastics technology concisely and comprehensively. The *Taschenbuch der Kunststoff-Additive* (R. Gächter, H. Müller, Carl Hanser Verlag; English edition: *Plastics Additives Handbook*) is a comprehensive reference on plastics additives and covers stabilizers, fillers, reinforcements and colourants for thermoplastics. It further contains a suppliers' index and a trade-name directory.

The analysis of polymers is described in *Polymeranalytik* (M. Hoffmann *et al.*, Thieme Verlag) in two volumes. The book contains a survey of the physical methods of characterizing polymers.

Journals

The following chapter concerns European journals that mainly publish polymer articles. However, it should be mentioned that articles of interest to polymer chemists can be found in other journals. For example *Nachrichten aus Chemie, Technik und Laboratorium* reviews the highlights in chemistry once a year; one chapter in this publication always concerns polymer chemistry. In order to have a complete survey of polymer chemistry one has to read the abstracting services; they usually cover journals with only a few polymer articles, as well as the specialist publications.

Kunststoffe is the oldest plastics journal in the world and provides technical information on plastics processing and applications in Europe with the largest proportion of exclusive articles coming from industry. *Kunststoffe* is the plastics publication most frequently cited in international technical literature. There is a bilingual edition of *Kunststoffe* that contains a complete section of English translations of all the original German articles. In 1987 there were four foreign-language editions of *Kunststoffe*, in Spanish (*Plásticos universales*) and even one in Russian and Chinese.

Die Makromolekulare Chemie is the leading European journal on macromolecular science. Every issue is composed of three sections: 1. Chemistry of Macromolecules; 2. Physical Chemistry of Macromolecules; 3. Physics of Macromolecules. *Die Makromolekulare Chemie* is an English-language journal although there are some contributions in German or French which have English summaries. *Die Makromolekulare Chemie — Rapid Communications* informs quickly about new ideas that may have great impact on macromolecular science. *Die Makromolekulare Chemie — Macromolecular Symposia* is published at least six times a year and contains presentations made at selected international symposia in the field of macromolecular chemistry.

Die Angewandte Makromolekulare Chemie is devoted to macromolecular substances and polymers of practical interest. It includes fundamental and applied research on synthesis, modification, characterization, structure and properties of these substances. *Die Angewandte Makromolekulare Chemie* is an English-language journal with some contributions in German or French.

Kautschuk und Gummi Kunststoffe is the official journal of the German Rubber Society and of the Rubber Technology Standardisation Committee of the German Institute for Standardisation. It appears monthly as a subscription journal and is internationally recognized by the rubber processing industry, by consumers of rubber and plastics as well as by the chemical industry responsible for the production of raw materials and by the relevant producers of machinery, plant and testing apparatus. *Kautschuk und Gummi Kunststoffe* reports in special columns, regularly and at short notice, on the latest developments and provides information on all specialist events taking place at home and abroad.

There are some other European polymer journals dealing mainly with applications that should be mentioned: *Revista de Plasticos Modernos* in Spanish; *Materie Plastiche ed Elastomeri* in Italian; *Plastiques Modernes et Elastomères* in French; *Swiss Plastics*, *Österreichische Kunststoff-Zeitschrift*, *Kunststoffberater*, *K-Plastic-Kautschuk-Zeitung* and *Plastverarbeiter* in German; *Colloid Polymer Science* in English.

Abstracting services

With the increasing number of patents and journals the importance of 'secondary literature' abstracting the primary literature has also increased. Beside *Chemical Abstracts* and *Derwent* which cover the whole field of chemistry, including polymer chemistry, some specific European abstracting services are of special interest to polymer chemists. Some abstracting services form the basis of data banks so that retrospective searches can be done.

The *Hochmolekularbericht* edited by Bayer AG in Leverkusen publishes abstracts of patents, conference papers and selected articles in journals including Russian and Japanese. The *Hochmolekularbericht* contains some 18,000 abstracts per year: about 11,000 abstracts of patents and 7,000 abstracts of articles selected from around 200 journals. If necessary the abstracts are illustrated by formulae. The abstracts are classified in seven main groups which are further subdivided under some 70 headings. Main groups include reviews, preparation of polymers, additives, applications and physical chemistry. The *Hochmolekularbericht* is published in German twice a month.

The literature reference service *Kunststoffe Kautschuk Fasern (Plastics Rubbers Fibres)* is edited by the Deutsches Kunststoff-Institut (DKI) Darmstadt and by the Fachinformationszentrum (FIZ) Chemie GmbH, Berlin. The DKI continuously explores the literature in the areas of plastics, synthetic fibres, elastomers and the physics and chemistry of polymers. Approximately 200 German and foreign journals, conference reports, monographs and German standards are scanned. All publications concerning the production, processing, applications and properties of macromolecular materials are selected, referenced, classified and indexed according to the DKI Thesaurus and stored in the information file *Kunststoffe Kautschuk Fasern*. Patents are not included. The information file is enlarged by about 12,000 literature references yearly. The abstract journal appears monthly with approximately 1,000 abstracts of articles per issue. The content is classified according to objective criteria according to the DKI classification. Cross references simplify the retrieval of desired articles.

Abstracts of patents and articles from journals are published monthly in French *Pascal Folio. Part 24: Polymères, Peintures, Bois* and the *Bulletin de Documentation du Centre d'Etude des Matières Plastiques, Paris*.

The *Kunststoff Information, Bad Homburg* informs weekly on companies, new products, prices and applications of plastics.

Conferences and meetings

Conferences and meetings are an opportunity for scientists, students, manufacturers etc. to exchange ideas with colleagues. A list of important meetings including polymer symposia is given by the Gesellschaft Deutscher Chemiker (GDCh) in Frankfurt. Publications of international symposia, especially selections from IUPAC symposia in the field of macromolecular chemistry, are found in *Die Makromolekulare Chemie — Macromolecular Symposia*.

One of the most important meetings for polymer producers and users is the International Plastics and Rubber Fair which takes place in Düsseldorf about every three years. The first was held in 1952 and the latest in 1986 was attended by 220,000 visitors from

111 countries. The trade fair informs about raw materials, auxiliaries, semi-finished and finished products, machinery, moulds and accessories. Detailed reports on the trade fair are published in the journal *Kunststoffe*.

A regular meeting of polymer scientists takes place in Freiburg every year in spring. The Makromolekulares Kolloquium is organized by the Institute for Macromolecular Chemistry in Freiburg. The papers presented at the meeting are not published anywhere else.

The Fachgruppentagung GDCh-Fachgruppe Makromolekulare Chemie takes place regularly in Bad Nauheim. The meeting discusses a main theme, e.g. new polymers or biopolymers.

Data bases and polymer search

The English sources of information like *Derwent* and *Chemical Abstracts* are of great importance but there are also other interesting European data bases. The *KKF (Kunststoffe Kauschuk Fasern)* data bank is produced by the DKI in Darmstadt and FIZ Chemie in Berlin and is accessible on the Host STN International. It contains references to articles which have been published since 1973. Sources, key words and the classification of articles are given. Since 1979 abstracts have also been included. The DKI performs retrospective literature searches through all the stored information. There are also given profile services. According to profiles of interests given by the customer, articles of newly published research are selected which match these profiles. The profile service appears monthly. The content is arranged according to the customer's wishes. For example, classification codes, controlled terms, supplementary terms, editors, language etc. may form the basis of a search. The controlled terms are based on a special Thesaurus which can be displayed online.

The *Polymat* data base contains numerical data and properties of thermoplastics, thermosets and casting resins. In its completed form *Polymat* will cover 6,000 materials, classes and types of plastics according to producer's trade names, 200 characteristics and properties in total and 60–80 characteristics and properties for each material. *Polymat* is designed to assist engineers as well as manufacturers in reviewing the available plastics and selecting materials for use. Furthermore, the data base is directed at all those working in the plastics field such as producers and organizations doing research and standardisation work. The data base is accessible online supported by a retrieval system and additional help functions. A manual, a keyword list and a learning programme are available. At a later stage a menu-driven user guide will be implemented. The data base is implemented by FIZ Chemie on the Host INKADATA and is accessible online via packet switching networks such as Datex P, Tymnet or Telenet. Requests for searches can also be directed to the DKI, FIZ Chemie or other information services. In the future an English version of *Polymat* will also be offered.

The *Kunststoff-Datenbank* is offered by the Kunststoff-Zentrale in Zürich. The *Kunststoff-Datenbank* contains about 10,000 entries covering some 430 companies and concerning raw materials, apparatus, trade names, etc. The *Kunststoff-Datenbank* provides contacts between producers and users.

There are some other European data bases that give general information including polymer data. *DITR* contains standards and technical rules and is accessible on FIZ Technik, Frankfurt. *PATDPA* is a German patent data base. It contains 900,000 entries, all in German, and includes references to German science and technology patents, patent applications and utility models from 1973 to the present. It is accessible on STN International. Similar information is given by *PATOS* which is produced by the Bertelsmann Informations-Service GmbH and the WILA-Verlag.

General information about European patents can be obtained from data bases produced by the Institut National de la Proprieté Industrielle (INPI), Paris. They are accessible on Questel. *INPI-1* contains patents applied for and published in France since 1969; *INPI-2* informs on all European patents, published applications and *EURO-PCT* since 1978; *INPI-3* contains patent documents of major industrialised countries since 1969 and *INPI-MARQUES* provides information on trademarks registered and published in France since 1981.

Titus covers mainly patents and journals in the textile field and is produced by the Institut Textile de France. It is accessible on Questel.

There are also general sources for retrospective searches. They are done, for example, by FIZ Chemie in Berlin or by the European Patent Office at the Hague.

CHAPTER EIGHTEEN

The plastics industry in the countries of the CMEA

L. SCHNEIDER

The proportion of total world production of plastics attributed to the CMEA countries was 12.2% in 1985. This proportion was a considerable increase compared with 1970, when it was only 9%. Development of the petrochemical industry, based on Soviet mineral oil, began relatively late in the countries of the CMEA. Agreements concluded amongst socialist countries, aimed at establishing capacities and co-operation, were of great importance. These agreements made it possible for the smaller countries of the CMEA to build up their own capacities in an economical way.

The growth of the chemical, and within this the petrochemical industry, from 1955–75, was an important development within the CMEA countries. Inexpensive Soviet mineral oil was the raw material for the plastics industry. The first oil crisis did not basically affect the situation, although the second oil crisis brought a considerable change. The Soviet Union was compelled to exploit its mineral oil resources in the Far East and in Siberia. The so-called 'Bucharest price-principle' setting the price of the mineral oil on the basis of the previous five years, contributed to the increased price of the Soviet mineral oil. The capacities of the plastic industry were greater than necessary to cover the domestic demands and were established with the objective of exporting, but due to the increased mineral oil prices it was no longer an economically viable proposition.

Plastics production of the CMEA countries was 9,140,000 tons in 1985, an increase of 12.9% over the previous year. The biggest producer was the Soviet Union with a production of 4,000,000 tons. The German Democratic Republic (GDR) and Czechoslovakia each produced more than 1,000,000 tons of plastics.

Production of the CMEA countries is targeted on standard

plastic products, the largest amounts being of PVC and polyethylene. Individual CMEA countries are characterized by the following.

Plastics production in **Bulgaria** is not of great importance. The first PVC plant with 12,000 tons per annum (p.a.) capacity was built comparatively recently in the mid-1960s in Devnja, in co-operation with the GDR and Czechoslovakia. 80,000 tons of polypropylene have been produced since 1981. However, production of ethylene oxide, ethylene glycol and polyurethane, in order to cover domestic demands, is not planned. Per capita consumption in Bulgaria in 1978 was 19 kg, currently increasing to 48 kg. In order to cope with the demand, large quantities of plastics are being imported from the Soviet Union and other socialist countries.

German Democratic Republic plastics production has been established for many years, annual production in 1985 being almost 1,250,000 tons. The large number of products includes low-pressure polyethylene, polystyrene, ABS and SAN in Schkopau and polyurethane in Schwarzheiden. In Schkopau, construction of a 30,000 tons p.a. capacity low-pressure polyethylene plant is planned. Per capita plastic consumption in the GDR is large (only Czechoslovakia is higher) at 63 kg. Exports from the GDR are considerable, particularly of artificial rubber, polyethylene and polyurethane. Most imports come from the Federal Republic of Germany (FRG).

Due to financial problems **Poland** has not developed the plastics industry. The present PVC capacity is 310,000 tons p.a., low-pressure polyethylene capacity is 141,000 tons p.a.and in addition 20,000 tons of polystyrene are being produced. There is a big shortfall in low-pressure polyethylene, polypropylene and polystyrene. There is no possibility however, of meeting this demand by increasing imports. The per capita consumption is very low at 18 kg.

The plastics industry in **Romania** has developed significantly between 1971–75, with an average annual increase of 16%. Since then economic and financial problems have slowed down development. Mineral oil production decreased sharply from 30,000,000 to 11,500,000 tons. Importation of mineral oil became necessary, but shortage of free currency caused difficulties. The Soviet Union delivers relatively small quantities and of a type from which it is not feasible to produce added value products which could meet Western demands. Under such circumstances there is no serious intention to develop the plastics industry. The current per capita consumption of 29 kg is rather low compared with the socialist countries, rising from 25 kg in 1978. This increase in consumption is the smallest among the socialist countries.

Although the planned production of the **Soviet Union** for 1985 was 6,200,000 tons actual production amounted to 4,400,000 tons. This was in spite of the fact that between 1981–85 several new plastics plants were brought on-stream. These plants comprised 200,000 tons low-pressure polyethylene, 120,000 tons high-pressure polyethylene, 100,000 tons polypropylene, and 250,000 tons PVC capacity. The per capita consumption of plastics (considering the huge territorial size of the country and the vast population) is also very low, 14 kg, rising from 11 kg in 1978–80.

The plastics industry of **Czechoslovakia** is one of the most developed among the socialist countries. All four high-volume standard plastic products, PVC, polyethylene, polypropylene and polystyrene are being produced. Low-pressure polyethylene capacity has been increased from 80,000 to 115,000 tons p.a. The older low-pressure plastics plant has been modernized by the Japanese Nisso Iwai and Kobe Steel Works. Per capita plastics consumption in Czechoslovakia of 67 kg is the largest within the socialist countries and over the past seven years has increased significantly by 40%.

In 1950 plastics production in **Hungary** was a mere 1,000 tons p.a. Nowadays it amounts to 423,000 tons p.a. The first plastics plant was built in 1962 with a capacity of 6,000 tons p.a. The foundation of the Hungarian petrochemical industry was based on the Soviet-Hungarian olefin agreement and was implemented within the framework of the Central Petrochemical Development Programme. Processing of raw materials produced in the Olefin Works was first carried out in 1979 in the PVC plant, using Shin Etsu technology. 180,000 tons of PVC were produced in 1979. The first polypropylene plant of 40,000 tons p.a. capacity was commissioned in 1978. In 1983 a new plant was brought into production with the same capacity. Until 1980 production of polyethylene was 50,000 tons p.a. The new linear polyethylene plant considerably increased the production of this type of material. In 1986 86% of plastics produced consisted of polyolefins and PVC. The per capita consumption was 41.5 kg in 1986, a considerable increase over previous years, due to the further development of these production and processing facilities. Hungary has significant exports, compared with production, to both socialist and non-socialist countries. Imports mainly consist of special and technical grade plastics products.

Organizations

Potentially useful organizations in the CMEA countries are listed below. Since commercial organizations and enterprises, publishing houses, technical and scientific libraries, universities and academic institutes are listed, it will be appreciated that as their activities are different so are the services available from them. Consequently their organizational structures are of different character as well. Nearly all of them are in the possession of computer facilities and some of them offer online services.

For example, activities of the Centre for Scientific and Technical Information, 125252 Moscow, Kuusinena, 216, include online services and access to various data bases in several fields of science and technology, special publications and periodicals, organization of seminars, conferences, preparation of compilations, studies, world-wide patent information, special translations, industrial catalogues, scientific and technical films, etc. Orders should be sent to: Vneshtekhnika, 119034 Moscow, Starokoniushenny per.6: Telex, 411418 'Molot'.

National Technical Information Centre and Library, H-1428 Budapest POB.12 Hungary: Telex, 22–4944 omikk h. The information services range from processing, publishing and dissemination of scientific, technical and economic information to offering online services and access to various foreign data bases, translations in the field of science, techniques and economies, organisation of conferences, meetings, symposia and presentations, reprographic services, direct mail, printing and distribution, publishing of conference proceedings, software products and services.

The enterprises mentioned in the list of addresses are mainly well informed in their own subject and especially in the products they manufacture. However, the laboratories attached to them may undertake research or may have achieved outstanding scientific results or elaborate new technologies.

If scientific or industrial co-operation is being planned it is advisable to contact the foreign trade enterprise of the particular country. More than one organization should be contacted simultaneously in order to obtain the necessary information.

Bulgaria

Centre for Research in Chemistry, Central Laboratory of Polymers, 1040 Sofia, 7 Noemvri 1.

Union of Chemistry and the Chemical Industry, 1000 Sofia, Rakowski 108.

Higher Institute of Chemical Technology, 8010 Burgas.

Higher Institute of Chemical Technology, Plastics Technology, 1156 Sofia, K. Ohridski 8.

Hozkombinat, Vidin.

Czechoslovakia

Forschungsinstitut 'Rubber and Plastics' für Gummi und Plasttechnologie, 76422 Gottwaldov 4.
Technische Gummi und Kunststoffwerke, Gottwaldov.
Research Institute for Petrochemistry, Dimitrovova 2/4, 97101 Prievidza.
VUTECHP Forschungsinstitut für Ökonomie der Chemischen Industrie, 11371 Praha 2, Stepanska 15.
PLASTIKA, 94953 Nitra.
East-Bohemian Works 'SYNTHESIA', 53217 Pardubice semtin.
UNICHEM, 53206 Pardubice, Nám. Budovateln 1458.
Chemische Werke der Tschechoslovakisch-Sowjetischen Freundschaft ZALUZI, v.Krusnych horach, Kraj Most.
CHEMOPETROL Chemische Werke, 43670 Litvinov, 18000 Praha 8, Liben Trojska 13 a.
Výzkumný Ústav Makromolekularni Chemie /SVUM/, Brno-Tkalcovaska 2.
Institut Ipolana, Neratovice.
Výzkumný Ústav Spracovania Aplikacia Plastickych Latok, Výskumna a vývojoda organizacia, 95037 Nitra.
Výskumný Ustav Gumarenske a Plastikovske Technologie, Gottwaldov.
TECHNOPLAST, 76811 Chropyne, Gottwaldov.
Povaeske Chemiché Zavody, 01034 Zilina.
Slovchema, Drienova 24, 82958 Bratislava.
Plastimat Kutelikova, 46078 Literei 6.
Slieroceske Chemiché Zavody, Lovisice.
Chemische Werke 'W Piech', Novaky 97271.
Technomat, Rybna 9, 11336 Praha 1.
Skloplast, Strojarenska 1, 91799 Trnava.
Institute of Chemical Processing Foundation, 165 02 Prague 6, Suchdol Rozvlová 135.
Institute of Macromolecular Chemistry, 162 06 Prague 6, Heyrovského nám.2.
Institute of Organic Chemistry and Biochemistry, 166 10 Prague 6, Flemingovo nám.2.
Czechoslovak Chemical Society, 118 29 Prague 1, Hrádcanské nám.12.
Centre of Chemical Research, 842 38 Bratislava, Dúbrovská cesta 9.
Institute of Polymers, 842 36 Bratislava, Dubravská cesta 9.
Slovak Chemical Society, 811 01 Bratislava, Gorkého 13.
Prague College of Chemical Technology (Faculty of Polymer Technology), 166 28 Prague 6, Suchbátarova 5.

College of Chemical Technology in Pardubice (Technology of Plastics), 532 10 Pardubice, Leninovo nám.565.
Slovak Technical University in Bratislava (Faculty of Chemical Engineering), 880 43 Bratislava, Gottwaldovo nám.17.

German Democratic Republic

VEB Chemische Werke BUNA, 4212 Schkopau.
VEB LEUNA Werke 'Walter Ulbricht', 4220 Leuna.
VEB Kombinat Plast und Elastverarbeitung Berlin, Aussenstelle Halle, 401 Halle (Saal), Grosse Ulrich Str.16.
VEB Ammendorfer Plastwerk, 4011 Halle, Schacht Str.11.
VEB Orbitaplast, 4371 Weissandt Golzau.
VEB Plastverarbeitungswerk, Staaken 1546.
VEB Chemiewerk Greiz-Dölau, 660 Greiz-Dölau.
VEB Eilenburger Chemie-Werk, 7280 Eilenburg, Ziegelstrasse 2.
Institut für Polymerenchemie, 1530 Teltow-Seehof, Knatstrasse 55.
Institut für Technologie der Polymere, 8010 Dresden, Hohe Str.6.
Institut für Chemische Technologie, 1199 Berlin, Adlershof, Rudower Chaussee 5.
Chemische Gesellschaft der DDR, 1080 Berlin, Clara Zetkin Str. 105.
Humboldt Universität zu Berlin (Department of Chemistry), 1086 Berlin, Unter den Linden 6.
Techniksche Hochschule Karl-Marx-Stadt (Department of Chemistry and Materials), 9001 Karl-Marx-Stadt, Str.der Nationen 62.

Hungary

Müanyagipari Kutató Intézet, H-1143, Hungaria krt.114.
Müszaki Kémiai Kutató Intézet, H-1111 Budapest, Budafoki uk 47.
Szerves Vegyipari Fejlesztõ Közös Vállalat, H-1085 Budapest, Stáhly u. 13.
Anyagmozgatási és Csomagolási Intézet, H-1085 Budapest, Rigó u.3.
Gumiipari Kutató Intézet, H-1087 Budapest, Kerepesi ut 17.
Tiszai Vegyi Kombinát, H-3580 Leninváros, P.O.B.20.
Borsodi Vegyi Kombinát, H-3700 Kazincbarcika, P.O.B.430.
Északmagyarországi Vegyimüvek, H-3792 Sajóbábony Gyártelep.
Országos Kõolaj és Gázipari Tröszt, H-1117 Budapest, Schönherz Zoltán u.18.

Országos Müszaki Fejlesztési Bizottság, H-1052 Budapest, Martinelli tér 8.
Graboplast, H-9023 Gyõr, Fehérvári ut 16.
Nitrokémiai Ipartelepek, H-8184 Füzfõgyártelep, P.O.B. 45.
Gépipari és Automatizálási Fõiskola, H-6000 Kecskemét, Izsáki ut 10.
Bõr-és Cipõipari Kutató-Fejlesztõ Vállalat, H-1047 Budapest, Paksi József u.43.
CHEMOLIMPEX, H-1805 Budapest, Deák Ferenc u.7–9.
Data Bank of Plastics Materials, Research Institute of the Plastics Institute, Budapest (see *Muanyag es Gumi*, **22**, (11), 1985, 317–22).
MTA Központi Kémiai Kutató Intézet (Central Research Institute for Chemistry of the Hungarian Academy of Sciences), 1025 Budapest, Pusztaszeri ut 59/67.
Budapesti Müszaki Egyetem (Technical University of Budapest), Faculty of Chemical Engineering (Plastics and Rubber), H-1521 Budapest, Müegyetem rkp.3.

Poland

Zaklad Inzynierii Chemicznej PAN (Institute of Chemical Engineering), 44–100 Gliwice, ul.Baltycka 5.
Instytut Chemii Przemislovej — Polymer and Plastic Technology (Institute of Industrial Chemistry), 01–793 Warsaw, ul.Rydygiera 8.
Instytut Przemyslu Tworzyw i Farb (Institute of Plastic and Paint Industry), 44–101 Gliwice, ul. Chorzowska 50.
Instytut Wlokien Chemicznych (Institute of Chemical Fibres), 90-570 Lódz, ul.Sklodowskiej-Curie 19–27.
Universytet Warszawski (University of Warsaw), Faculty of Chemistry, 00–325 Warsaw, Krakowsie Przedmieście 26–28.
Politechnika Gdanska (Gdansk Technical University), Polymer Chemistry and Technology, 80–233 Gdansk, ul. Majakowskiego 11/12.
Politechnika Szczecinsja (Szczecin Technical University), Polymer Technology and Engineering, 70–310 Szczecin, ul. Piastów 17.

Romania

Institutul Central de Chimie (Central Institute of Chemistry), 77208 Bucharest, Str.12.Splailul Independentei 202.

Universitatea Cluj-Napoca (Faculty of Chemistry), 3400 Cluj-Napoca, Str.M.Kogalniceanu 1.
Institutul Politehnik 'Gh.Gheorghiu-Dej' (Faculty of Equipment and Chemical Process Engineering), Bucharest, Splaiul Independentei 313.
Institutul Politehnic 'Gheorghe Asachi' (Faculty of Chemical Technology), Iasi, Calea 23 August 22.
Institutul Politehnic 'Traian Vuia' Timişoara (Faculty of Chemical Technology), Timişoara Bd.30 Decembrie 2.

Soviet Union

NORPLAST, 111394 Moscow, Ul.Perovskaya 66.
Institute of Chemistry, Gorky, Pochtamt, Ul.Tropinina 49.
Kornakov, N.S. Institute of General and Organic Chemistry, Moscow, Leninsky Pr.31.
Institute of Chemistry, Vladivostok, Pr.100-letia.
Institute of Chemistry, Sverdlovsk region, Pervomaiskaya 91.
Institute of General and Inorganic Chemistry, 375051 Erevan, Ul.Fioletova 10.
Institute of Chemistry, 200026 Tallinn, Akadeemia tee 15.
Institute of Mechanics of Polymer Compounds, Riga, Ul.Aizkarukles 23.
Institute of Chemistry and Chemical Technology, Vilnius, K.Poželos 48.
Institute of Chemistry, Kishinev, Ul.Akademicheskaya 3.
Institute of General and Inorganic Chemistry, Kiev, Or.Akademika Palladina 32/34.
Mendeleyev, D.I. All-Union Chemical Society, Moscow, Krivokolenny per.12.
All-Union Research and Project Institute of Man-made Fibres, Moscow region, Mytishchi, Ul.Kolontsova 5.
Gorky N.I. Lobachevsky State University, Chemical Institute, 603091 Gorky, Gagarina 23.
Kharkov A.M. Gorky State University, Institute of Chemistry, 310077 Kharkov, Pl.Dzerzhinskogo 4, Ukraine.
Leningrad A. Zhdanov State University, Faculty of Chemistry, 199164 Leningrad, Universitetskaya nab.7/9.
Moscow M.V. Lomonosov State University, Faculty of Chemistry, 119808 Moscow, Leninskie Gory.
Vilnius V. Kapsukas State University, Faculty of Chemistry, 232734 Vilnius, Universiteto 3. Lithuanian SSR.

Moscow M.V. Lomonosov Institute of Fine Chemical, Technology Polymer Processing Technology, Moscow, Malaya Pirogovskaya Ul.1.
Leningrad Lensoviet Technological Institute, 198013 Leningrad, Zagorodny pr.49.
Plasticheskije Massy, Redakciya: 129110, Moscow, Ul. Gilyarovskogo 39.
NPO 'Plastmassy', 111112 Moscow, Perovskij Pr.35.
NPO 'Plastik', 121059 Bereshkovskaya Nad.20.
'Himavtomatika' 300023 Tula, Ul. Bolgina 94.
NII Himii i Technologii Polimerov im. Akademika Bakargina, 606006 Dzerzhinsk, Gordovskoi Obl.

Yugoslavia

Union of Engineers and Technicians of Yugoslavia, Union of Chemists and Technologists of Yugoslavia, Belgrade, Kneza Milosa 9/11.
Serbian Chemical Society, 11000 Belgrade, Karnegieva 4.
Sveuciliste u Zagrebu (University of Zagreb), Faculty of Technology, 41000 Zagreb, Trg Marsala Tita 14.

CHAPTER NINETEEN

The Americas — North and South

J. R. LAWRENCE

Information on the rapidly developing and changing plastics industry in North and South America comes from a variety of sources both within and outside of the industry. This chapter presents a compilation of publications available from these information sources. These references are primarily from a publication entitled *International Sources of Plastics Information* originally prepared by The Society of the Plastics Industry in 1983. No attempt is made to identify the 'online' computer sources of information, such as that offered by The American Chemical Society's Chemical Abstracts Service (CAS) data base which are covered elsewhere in the volume.

It must be understood that the dynamics of the industry are such that new information is constantly evolving. Therefore the studies and reports cited herein may not represent the most current information available. When making inquiries to the identified organizations for information it will be important to request information on the *latest* version of the cited study and other reports of a similar nature.

Each of the information sources views the industry from a unique perspective which must be understood in interpreting the resulting data. The principal information comes from the following sources:

1. Trade and Professional Associations: These organisations gather information, through their membership, on the specific activities with which they are directly involved. This results in some of the most accurate information on production of basic materials; sales by market categories; and management concerns in areas such as labour rates, financial data, salaries, etc.

2. Trade Publications: Industry publications provide broad perspectives on industry progress from a news standpoint. This information tends to emphasize major trends in materials, processing techniques, competition, and obstacles to progress.
3. Government or Quasi-government Agencies: Government agencies, such as the US Department of Commerce, organize data in terms of general industry statistics and indices which serve as business barometers.
4. Consultants and Financial Institutions: Information from these independent organisations tends to probe the advanced and specialised areas of the industry requiring interpretation not readily available to the above organisations.
5. General Textbooks: These publications are primarily tutorial and involve basic instructional information of long standing value.
6. Plastics Industry Manufacturers: A limited number of companies within the industry publish 'house organs' which highlight progress in the development of their own materials and products.
7. South America — Trade Associations, Trade Publications, and Consultants: These references cover activities in the principal countries in South America where there are significant plastics manufacturing activities.

These references are grouped in sections by source to categorize material of a somewhat similar nature as described above. Each listing shows the following: Report No.; Title of Report; Frequency of Publication; Publisher of material; Address of publisher; Price of publication, if known; Brief abstract of material. Following the listing of these 127 literature references, there is a cross-referenced index (Table 19.1) of the material which will assist in identifying sources of particular types of information. This index can be used as a general guide to identify sources for specific types of information. It should not be considered as a complete index of the full contents of these reports.

Trade and professional associations:

1; *Annuario de la industria Quimica*; Annual; ANIQ; Providencia No. 1118; Mexico 12, D.F. Mexico; $ — unknown; Production and sales data on Mexican plastics industry.
2; *Facts and Figures of the U.S. Plastics Industry*; Annual; The Society of the Plastics Industry, Inc., Statistical Department; 1275 K Street, N.W., Suite 400, Washington, D.C. 20005; $90.00; Details production, sales, markets, capacities for major plastics. Background data include producer price indexes, feedstock data, finished products, additives, machinery, and major markets.

3; *Financial and Operating Ratios Survey*; Annual; The Society of the Plastics Industry, Inc., Statistical Department; 1275 K Street, N.W., Suite 400, Washington, D.C. 20005; $100.00; Measures the significant financial operating ratios affecting plastics processing firms such as assets and liabilities, net sales, pretax and after tax income, cost of sales, overhead, depreciation, inventory, profits, and administrative expenses.

4; *Labor Survey*; Annual; The Society of the Plastics Industry, Inc., Statistical Department; 1275 K Street, N.W., Suite 400, Washington, D.C. 20005; $100.00; Statistics for plastics processing companies including hourly wages, fringe benefits, vacations, overtime pay, holidays, shifts, bonuses, and turnover.

5; Plastics: A.D. 2000 — production and use through the turn of the century; 1987; The Society of the Plastics Industry, Inc., 1275 K Street, N.W., Suite 400, Washington, D.C. 20005; $ — unknown; Predicts growth in the use of plastics through the year 2000 and details growth by resin, end-use market.

6; *Salary Survey*; Periodic; The Society of the Plastics Industry, Inc., Statistical Department; 1275 K Street, N.W., Suite 400, Washington, D.C. 20005; $100.00; Measures the salary and total compensation for employees of plastics processing companies. Data are shown for salaried employees in manufacturing, marketing, finance, engineering, personnel and purchasing.

7; *SPE Membership Roster — SPE Journal*; Annual; Society of Plastics Engineers; 14 Fairfield Dr., Brookfield, 06805; $ — unknown; Roster lists approximately 13,000 names, addresses, company affiliations and job titles. Provides a list of Society chapters.

8; *SPI Membership Directory and Buyers Guide*; Annual; The Society of the Plastics Industry, Inc., 1275 K Street, N.W., Suite 400, Washington, D.C. 20005; $100.00; Consists of an alphabetic-geographic listing of 1,900 companies. Subject index.

9; *Statistical Report on Thermoplastic and Thermosetting Resins*; Monthly; Ernst and Whinney (Published for SPI); Trade Association Services Dept., New York, 10022; $400/year; Production, sales, and captive use data on major plastics, showing current month, same month a year earlier, and year-to-date comparisons with percentage change.

Trade publications

10; *Adhesives Age*; Monthly; Communication Channels, Inc., 6285 Barfield Rd., Atlanta, Georgia 30328; $19/year; Trade publication covering adhesive materials and their applications.

11; *Adhesives Red Book*; Annual; Palmerton Publishing Co., Inc., 461 8th Ave., New York, 10001; $ — unknown; Lists manufacturers and suppliers in the adhesives industry, including consultants, laboratories.
12; *Canadian Plastics*; Monthly; Southam Communications Ltd.; 1450 Don Mills Road; Don Mills, Ontario M3B 2X7, Canada; $21/year; Trade publication covering Canadian plastic materials and their applications.
13; *Canadian Plastics Directory and Buyers' Guide*; Annual; Southam Communications Ltd., 1450 Don Mills Road; Don Mills, Ontario M3B 2X7, Canada; $25.00; Canadian plastics magazine.
14; *Chemical and Engineering News*; Weekly; American Chemical Society; 1155 16th Street, N.W., Washington, D.C. 20036; $24/year; Covers news and developments in chemicals and their markets.
15; *Chemical Engineering*; Fortnightly; McGraw-Hill Publications Co.; 1221 Ave. of the Americas, New York, 10020; $20/year; Covers developments in chemical engineering and markets for chemicals and chemical technology.
16; *Chemical Marketing Reporter*; Weekly; Schnell Publishing Co.; 100 Church St., New York, 10007; $32/year; Covers developments in all markets for chemicals.
17; *Chemical Week*; Weekly; McGraw-Hill Publications Co.; 1221 Ave. of the Americas, New York, 10020; $22/year; Covers developments in chemicals, chemical markets, and the chemical industry.
18; Dynamics of the plastics marketplace (4th edn); 1980; Cahners Publishing Co.; 275 Washington St., Newton, 02158; $95; Documents the purchasing process for industrial products, materials and equipment. Answers six questions: frequency of purchase influence; dollar volume of purchases; types of products influenced; areas of purchasing influence; dependence on sales representatives; role of information sources.
19; *Hule Mexicano y Plasticos*; Monthly; Juan Solorzano Gomez; Filomeno Mata 13–11, Mexico 1, D.F. Mexico; Mex. $150/year; Covers plastic industry activities in Mexico.
20; *Modern Plastics*; Monthly; McGraw-Hill Publications Co.; 1221 Ave. of the Americas, New York, 10020; $24/year; Trade publication covering plastics materials and their applications.
21; *Modern Plastics Encyclopedia*; Annual; McGraw-Hill Publishing Co.; 1221 Ave. of the Americas, New York, 10020; Included with prices of subscription to *Modern Plastics*. Presents entries on more than 4000 companies of plastic materials, machinery, and other services.
22; *Plastics Design Forum*; Bimonthly; Harcourt Brace Jovanovitch;

New York, $12/year (free to qualified personnel); Trade publications specializing in plastics product design technology.
23; *Plastics Design and Processing*; Monthly; Lake Publishing Corp.; 700 Peterson Rd., Libertyville, Illinois; $25/year; Trade publication covering plastics materials, processing and design of plastics products.
24; *Plastics Focus*; Weekly; Plastics Focus; 1601 Third Ave., Room 11F/W, New York, New York 10028; $120/year; A weekly interpretive newsletter covering developments in the plastics industry.
25; *Plastics Machinery and Equipment*; Monthly; Harcourt Brace Jovanovitch; New York, New York; Free to qualified personnel; Trade publication reporting on specialised machinery and equipment for manufacture of plastics products.
26; *Plastics Technology*; Monthly; Bill Communications, Inc.; 633 Third Ave., New York, 10017; $15/year; trade publication covering plastics materials and their applications.
27;*Plastics World*; Monthly; Cahners Publishing Co.; 275 Washington St., Newton, 02158; $30/year; Trade publication covering plastics materials and their applications.
28; *Plastics World Reference File*; Annual; Cahners Publishing Co.; 275 Washington St., Newton, Massachusetts, 02158; $15.00; Lists over 3,000 manufacturers and suppliers in 700 product categories.
29; *Urethane Plastics and Products*; Monthly; Technomic Publishing Co., Inc.; 851 New Holland Ave., P.O. Box 3535, Lancaster, Pennsylvania 17604; $45/year; Covers developments in urethane polymers and their markets.
30; *Western Plastics Directory/Yearbook*; Annual; Western Plastics News, Inc.; 1740 Colorado Ave., Santa Monica, California 09494; $ — unknown; Lists of manufacturers west of the Mississippi River.

Government or quasi-government agencies

31; *Directory of Plastics Processors and Producers in North Carolina*; Unknown; North Carolina State University Industrial Extension Service; Box 5506, Raleigh, North Carolina 27607; $4.00; Lists of North Carolina processing companies; product line services offered, key personnel.
32; *Directory in Plastics — Knowledgeable Government Personnel*; Periodic; U.S. Dept. of Commerce Rpot. No. 642574; Clearing House for Federal & Scientific Information, Springfield, Virginia 22151; $ — unknown; Material on all military and government agencies working in plastics research and development; engineers

engaged in developing and manufacturing; names and addresses and telephone numbers of installations and individuals.

33; *Imports of Benzenoid Chemicals and Products*; Annual; U.S. International Trade Commission; 1272 8th & E Streets, Washington, D.C. 20436; $ — unknown; Statistics on imports of benzenoid chemicals including intermediates, dyes and pigments, medicinals and pharmaceuticals, and plastics.

34; *Plastics Manufacturing Capabilities in Mississippi*; Periodic; Mississippi Research and Development Center; P.O. Drawer 2470, Jackson, Mississippi 39205; Free; Lists plastics manufacturing facilities in Mississippi.

35; *Plastics Processors and Fabricators in Georgia and Surrounding Areas*; Periodic; Georgia Institute of Technology, Industrial Development Division, Engineering Experimental Station, Basic Data Branch; Atlanta, Georgia 30332; $3.00; Lists plastics manufacturing facilities in Georgia and surrounding areas.

36; *Producer Price and Price Indexes*; Monthly; U.S. Dept. of Labor, Bureau of Labor Statistics, Superintendent of Documents, U.S. Gov't Printing Office; Washington, D.C. 2402; $20/year; Report on producer price movements including text, tables, and technical notes. An annual supplement contains monthly data for the calendar year, annual averages, and information on weights and changes in the sample.

37; Profiles of material supplier industries to the automotive manufacturers; 1981; U.S. Dept. of Transportation, #DOT-TSC-NHTSA-81-16, National Technical Information Service; Springfield, Virginia 22161; Free; Study of industries supplying materials to the automobile manufacturers, such as steel, castings, aluminium, glass, paint, and plastics.

38; *Synthetic Organic Chemicals*; Annual; U.S. International Trade Commission; 1183 8th & E Streets, Washington, D.C. 20436; $8.00; Production and sales data on all U.S. chemicals from 800 companies. Divided into 15 sections: tar and tar crudes; dyes; cyclic intermediaries; primary products from petroleum and natural gas for chemical conversion; organic pigments; medicinals; elastomers; plasticizers; surface-active agents; pesticides; plastic and resins; flavours and perfumes; rubber-processing chemicals; miscellaneous cyclic and acylic chemicals; and miscellaneous end-use chemicals.

39; *U.S. Industrial Outlook*; Annual; U.S. Dept. of Commerce, Superintendent of Documents, U.S. Government Printing Office; Washington, D.C. 20402; $ — unknown; Review of 200 industries giving current situation, outlook for next year and long-term prospects. Provides basic data on value of shipments, production workers, number of establishments, value of imports and exports.

Consultants and financial institutions

40; Acrylic polymers; February 1979; Skeist Laboratories, Inc.; 112 Naylon Ave., Livingston, New Jersey 07039; $4,750; Statistical data on acrylic monomers, polymers, and end-uses projected to 1983. Major suppliers with their market shares; industry structure; appraisal of trends; growth rates; opportunities and new product needs; impact of governmental regulations and policies.
41; Advanced composites, October 1982; Skeist Laboratories, Inc., 112 Naylon Ave., Livingston, New Jersey 07039; $6,000; Report covers industry review; polymers used as matrices; reinforcements; end-users; technology; new developments.
42; Advanced composites: commercial applications; October 1981; Business Comminications Co., Inc.; Box 2070 C, Stamford, Connecticut 06906; $950; Highlights economics and markets for composites with emphasis on technology and manufacturing methods. Companies; costs; forecasts.
43; Blow molding enters the eighties; December 1982; Business Communications Co., Inc.; Box 2070 C, Stamford, Connecticut 06906; $1,500; Analyses plastics as a bottle and can substitute. Applications by type, major manufacturers, container comparisons, newest innovations.
44; Competition, substitution between plastics; February 1981; Business Communications Co., Inc.; Box 2070 C, Stamford, Connecticut 06906; $950; Markets for HDPE, LDPE, PP, PS, PVA, ABS, phenolic and others are studied in depth. Projections to 1990. Government legislation and corporations involved included.
45; Covalent conductive polymers; August 1982; Business Communications Co., Inc.; Box 2070 C, Stamford, Connecticut 06906; $1,250; Analysis of current states of research and development and technology of conducting polymers and applications, markets, forecast.
46; *Directory of Chemical Producers*; annual; SRI International; 333 Ravenswood Ave., Menlo Park, California 94025; $400; Companies section lists 1,500 companies alphabetically by parent company name. Lists all subsidiaries, divisions and affiliates with addresses, plant locations and products. Product section and region section are cross-referenced.
47; Engineering Thermoplastics; June 1979; Skeist Laboratories, Inc.; 112 Naylon Ave., Livingston, New Jersey 07039; $6,000; Covers nylon, polycarbonate, acetals, thermoplastic polyesters, modified PPO, fluoropolymers, polysulphones, polyphenylene sulphide, ABS/SAN, acrylics, glass reinforced polypropylene.
48; Engineering thermoplastics; June 1982; Predicast, Inc.; 11001 Cedar Ave., Cleveland, Ohio 44106; $995; Analyses historic and

future (to 1995) use of high-performance thermoplastic polymers. Competition from lightweight materials and reinforced thermosets is reviewed. Major companies are profiled.
49; Engineering thermoplastic update; September 1982; Business Communications Co., Inc.; Box 2070 C, Stamford, Connecticut 06906; $1,500; Highlights new competing materials and compounds. Markets.
50; Ethylene copolymers; October 1982; Business Communications Co., Inc.; Box 2070 C, Stamford, Connecticut 06906; $1,250; Complete technical market analysis by product, applications for high volume and speciality materials. Includes companies, forecasts, application by type.
51; Fillers and extenders for plastics; May 1980; Business Communications Co., Inc.; Box 2070 C, Stamford, Connecticut 06906; $750; Data concerning glass bubbles, clay, silicates, waste products, applications, newest developments for fillers, extenders and reinforcements.
52; High temperature thermoplastics; October 1978; Business Communications Co., Inc.; Box 2070 C, Stamford, Connecticut 06906; $675; Study looks at 8 resin/compounds/users; current and future applications. Survey results included with performance comparisons.
53; Impact modifiers; March 1981; Business Communications Co., Inc.; Box 2070 C, Stamford, Connecticut 06906; $950; Flexible polyolefins, acrylics of all types; rubber, thermoplastic elastomers, ABS, polybutylenes complete with each other. Technology, price/property relationships, preferences, markets by resins and forecasts outlined.
54; Imports and exports of plastics; April 1981; Business Communications Co., Inc.; Box 2070 C, Stamford, Connecticut 06906; $800; Effects of oil prices, international competition, and other developments on imports and exports of plastics by country and material.
55; Kline guide to the chemical industry (4th edn); 1980; Charles H. Kline and Co., Inc.; 330 Passaic Ave., Fairfield, New Jersey, 07006; $147; Summary of the economics of the industry including trends and future outlook of shipments, profitability, expenditures, pricing, and raw materials. Directory of 525 companies with over $5 million in sales of chemical products.
56; Kline guide to the plastics industry; Triennial — 1985; Charles H. Kline and Co., Inc.; 330 Passaic Ave., Fairfield, New Jersey 077006; $155.00; Contains 500 leading firms and other organisations. An analysis of the industry is provided.
57; Linear low density polyethylene — opportunity or threat?; April 1980; Chem Systems Inc.; 303 Broadway, Tarrytown, New

York 10591; $8,000; Covers the economics of the available processes and the potential markets of LLDPE. Included is an analysis of the vapour phase process of Union Carbide and Napthachimie, the slurry process of Phillips and the solution processes of DuPont, Dow and Mitsui.

58; LLDPE vs. LDPE; August 1982; Predicast, Inc.; 11001 Cedar Ave., Cleveland, Ohio 44106; $11,000; Two volumes (film, non-film) on markets; acceptablility and limitations by end market; resin grades and forms; product optimisation by end market.

59; Multilayered film: new directions; September 1982; Business Communications Co., Inc.; Box 2070 C, Stamford, Connecticut 06906; $1,500; Study looks at all types of multilayered films and their markets, as well as technologies and practices in extrusion and converting.

60; New and speciality films; June 1978; Skeist Laboratories, Inc.; 112 Naylon Ave., Livingston, New Jersey 07039; $4,500; Trends, growth rates, opportunities, new product needs, industry structure and impact of government health, safety and environmental regulations.

61; Pallet load utilizing materials and equipment; 1981; Hull & Co.; 5 Oak St., Greenwick, Connecticut 06830; $5,000; Analyses properties, cost-performances parameters, producers, equipment technology trends, and issues affecting buyers and suppliers for plastic stretch film, shrink film, netting, strapping (steel and plastic) and adhesives.

62; Petrochemical feedstocks: how much? when?; August 1981; Business Communications Co., Inc.; Box 2070 C, Stamford, Connecticut 06906; $975; Examines the demand for major petrochemical products by type through 1990 and then beginning with an energy forecast by major energy sources in turn develops a supply/demand picture for the major feedstocks at two levels.

63; Petrochemical manufacturing and market trends; Irregular; Chem Systems Inc.; 303 Broadway; Tarrytown, New York 10038; $7,500.00/year; Data and analysis for both energy and feedstocks and basic petrochemicals supply/demand, costs and pricing. Includes crude oil, natural gas, natural gas liquids, coal and refinery products, ethylene, propylene, butylenes, butadiene, benzene, toluene.

64; Plastic additives; October 1980; Predicasts, Inc.; 11001 Cedar Ave., Cleveland, Ohio 44106; $775; Analyses and projects markets for plastics additives; includes forecasts to 1995.

65; Plastic additives market; January 1978; Frost and Sullivan, Inc.; 106 Fulton St., New York, New York 10038; $750; Forecasts through 1986 for 46 additives. Company market share by additive category, market by end-use. Pricing.

66; Plastic alloys; 1979; R.M. Kossoff and Associates, Inc.; 10 Rockefeller Plaza, New York, New York 10020; $2,900; Covers alloys of PVC, ABS, modified PPO, and others, including those based on engineering resins, rubber modified polyolefins. Markets, technology, producers, and technology sources.

67; Plastic bottles; October 1980; Predicast, Inc.; 11001 Cedar Ave., Cleveland, Ohio 44106; $325; Historic and projected data on plastic bottle production, resin use and end-use markets for food, beverage and non-food bottles. Reviews world producers, new technologies and government regulations.

68; Plastic film markets; November 1980; Predicasts, Inc.; 11001 Cedar Ave., Cleveland, Ohio 44106; $775; Examines packaging and other markets for major plastic films. Projections to 1995 for 30 markets including trash bags, shrink wrap, stretch wrap, shipping sacks, mag tapes, photo film, baked goods, frozen goods.

69; Plastic foams to 1995; March 1980; Predicasts, Inc.; 1101 Cedar Ave., Cleveland, Ohio 44106; $775; Analysis of plastic foams including rigid and flexible urethanes, PVC, polystyrene and others. 20 end-use markets are projected to 1985, 1990 and 1995.

70; Plastic processing machinery and markets; February 1978; Predicasts, Inc.; 11001 Cedar Ave., Cleveland, Ohio 44106; $775; Historical and projected to 1990 consumption of resins by process and machinery purchases by type. Analyses industry structure of equipment producers and end-users.

71; Plastic sheet vs. flat glass; June 1982; Business Communications Co., Inc.; Box 2070 C, Stamford, Connecticut 06906; $525; Competition takes place in the automotive and building/construction areas. Acrylic, polycarbonate and polyester sheet show cost advantages over glass. Chemical resistance and flammability are looked at.

72; Plastics and rubber machinery market; March 1978; Frost and Sullivan; 106 Fulton St., New York, New York 10038; $750; Analysis and forecasts through 1987 for all machinery types.

73; *Plastics Additives Marketing Guide*; Periodic; Technomic Publishing Co.; 851 New Holland Ave., Lancaster, Pennsylvania 17604; $ — unknown; Provides an overview of the industry with a directory listing producers, addresses, products, estimated sales.

74; Plastics forming; February 1980; Business Communications Co., Inc.; Box 2070 C, Stamford, Connecticut 06906; $850; Review of eight major resins included with market trends and limitations/advantages of new technology; companies involved.

75; Plastics in construction; September 1982; Predicast, Inc.; 11001 Cedar Ave., Cleveland, Ohio 44106; $995; Presents

historical and projected (to 1995) data for 30 end-use plastics markets, as well as comparative analysis of competitive materials.
76; Plastics international trade trends and prospects; May 1981; Business Communications Co., Inc.; Box 2070 C, Stamford, Connecticut 06906; $800; The international trade performance of the US plastics industry since 1973 is reviewed and prospects to 1985 are discussed. Country data, specific resins and materials, production, sales, exports/imports, companies, costs.
77; Plastics in food packaging; August 1980; Business Communications Co., Inc.; Box 2070 C, Stamford, Connecticut 06906; $975; Analyses the position of plastics versus competitive materials; growth and decline of major foods; food consumption habits of the public; change in packaging materials to provide longer shelf life.
78; Plastics in transportation: changing markets; October 1982; Business Communications Co., Inc.; Box 2070 C, Stamford, Connecticut 06906; $1,500; Report contrasts and compares consumption criteria for plastics in major transportation groups.
79; Plastics in business machines: how much? when?; February 1982; Business Communications Co., Inc.; Box 2070 C, Stamford, Connecticut 06906; $1,500; Reviews size of plastics in the business machine market and forecasts growth. Structural foam reinforced plastics, new grades reviewed.
80; Plastics vs. paper; November 1980; Business Communications Co., Inc.; Box 2070 C, Stamford, Connecticut 06906; $950; Price ranges and markets involved with details concerning bags/wrapping, printing/writing and sustitution data.
81; Plastics vs. paperboard; February 1981; Business Communications Co., Inc.; Box 2070 C, Stamford, Connecticut 06906; $950; Analyses important economic and market forces impinging on the paperboard/plastics interface. Report shows how, where, why and the extent of substitution.
82; Plastics vs. metals; August 1981; Business Communications Co., Inc.; Box 2070 C, Stamford, Connecticut 06906; $975; Reviews further opportunities for metal replacement in industrial and consumer applications.
83; Polyarylates; 1979; R.M. Kossoff and Associates, Inc.; 10 Rockefeller Plaza, New York, New York 10020; $1,900; Review of technology, economics and market potential of new transparent thermoplastic from Japan.
84; Polyurethanes, II; April 1983; Skeist Laboratories, Inc.; 112 Naylon Ave., Livingston, New Jersey 07039; $7,500; Analyses all raw materials and end uses for polyurethanes such as rigid and flexible foams, elastomers, coatings, adhesives, sealants. Government and association activities.

85; Polyurethane and other foams: opportunities; October 1978; Business Communications Co., Inc.; Box 2070 C, Stamford, Connecticut 06906; $675; Provides an in-depth and analytical survey of rigid and flexible polyurethane foam markets, materials and processing technology and manufacturing methods.
86; *Predicasts Forecasts*; Quarterly; Predicast, Inc.; 11001 Cedar Ave., Cleveland, Ohio 44106; $535/year; Contains one-line summaries of expert forecasts which have appeared in trade journals, business and financial publications, newspapers, government reports, and special studies. Each forecast shows subject, base period data, short and long range forecasts, source of the forecast, and projected annual growth rate.
87; Protective plastics packaging; 1982; Charles H. Kline and Co., Inc.; 330 Passaic Ave., Fairfield, New Jersey 07006; $50000; Analysis of the major cushioning and restraining products available, end-use applications, resins employed, competitive materials, major suppliers.
88; Reinforced plastics: markets and technological perspectives; March 1980; Business Communications Co., Inc.; Box 2070 C, Stamford, Connecticut 06906; $950; Report looks at markets by dollar and pounds, as well as resin and reinforcement consumption by type.
89: Reinforced plastics; March 1981; Predicasts, Inc.; 1101 Cedar Ave., Cleveland, Ohio 44106; $775; Historic and projected to 1995 sales of fibrous and non-fibrous reinforced plastics to 12 markets. Increasing role of specialised resins such as nylon and urethanes, and the outlook for composite and hybrid materials in each market.
90; Reinforcements and fillers for plastics 1980; 1980; Charles H. Kline and Co., Inc.; 330 Passaic Ave., Fairfield, New Jersey 07006; $12,000; study covers varied patterns of growth of consuming industries, large number of competing products and grades in numerous applications. Development of new products.
91; Resin switching; how much? how real?; August 1981; Business Communications Co., Inc.; Box 2070 C, Stamford, Connecticut 06906; $1,500; Reviews focus shift to the altered price relationship between resins and to new technologies and materials that are being developed to reduce the materials cost component of finished parts and to take advantage of charged price relationships in pitting one resin against another.
92; Resurging plastics pipe industry; April 1980; Business Communications Co., Inc.; Box 2070 C, Stamford, 06906; $750; Study looks at major markets; competing materials. Pipe markets by materials are listed with new technology outlined.

93; Rigid thermoplastic profile extrusion; 1981; Hull and Co.; 5 Oak St., Greenwich, Connecticut 06830; $10,000; Study covers full range of thermoplastic polymers used in rigid extrusions including blends, alloys and co-extruded materials. Included are ancillary materials used in blended compounds, equipment used in extrusions and fabrication, identification of major processors, specific profile market growth.

94; RIM breakthrough; February 1980; Business Communications Co., Inc.; Box 2070 C, Stamford, Connecticut 06906; $850; Early markets, product and market advances described in depth with developments in the evolving machine equipment arena and chemical technology in the processing field.

95; Saturated polyesters; March 1983; Skeist Laboratories, Inc.; 112 Naylon Ave., Livingston, New Jersy 07039; $6,500; Report covers monomers; industry review; and end-uses such as moulded plastics, film and sheet, coatings, adhesives, plasticizers, intermediaries.

96; Speciality polymer blends and alloys; 1982; Charles H. Kline and Co., Inc.; 330 Passaic Ave.; Fairfield, New Jersey 07006; $10,000; Analysis of the major polymer blends and alloys available, end-use applications, process technologies, competitive materials, and major suppliers.

97; Structural foam molding; October 1982; Business Communications Co., Inc.; Box 2070 C, Stamford, Connecticut 06906; $1,500; Report details consumption, provides moulder lists, count of machines in use by manufacturer, discusses new processes and trends in end-use markets, assets and liabilities of the process.

98; The compounding business; 1980; R.M. Kossoff and Associates, Inc.; 10 Rockefeller Plaza, New York, New York 10020; $3,400; Covers custom and proprietary compounding. Shows major competitors, apparent profitability and acquisition opportunities.

99; The Mexican plastics and resin industry; February 1981; William D. Gersumky, SP Latin America; 301 E. 69th St., New York, New York 10021; $650.000; Includes production, market and end-use data, demand projections to 1990. Lists major fabricators.

100; The outlook for the plastics industry through 1985; July 1980; Arthur D. Little, Inc.; 35 Acorn Pake, Cambridge, Massuchusetts 02140; $ — unknown; Identifies the largest and fastest growing markets. A discussion of changes in the plastics products fabrication business. Assessment of relative positions of leading producers.

101; The polystyrene industry 1978–1984; May 1979; Peter Sherwood Associates, Inc.; 60 East 42nd St., New York, New York 10165; $3,900; Analyses resin's 73 main market sectors.

More than 350 principal consumers are listed in their respective market segment. Their position in each market is indicated for almost all, and the actual polystyrene consumptions reported for nearly 250 converters.

102; The retortable pouch; November 1980; Business Communications Co., Inc.; Box 2070 C, Stamford, Connecticut 06906; $950; U.S. military needs are being met with this new method of canning and bagging.

103; Thermoforming 1978–1982; December 1978; Springborn Group; Hazardville Station; Enfield, Connecticut 06082; $2,250; Current status of thermoforming and examination of new forming processes and materials and their impact in the consumer and industrial marketplace.

104; Thermoplastics to 1995; March 1982; Predicast, Inc.; 11001 Cedar Ave., Cleveland, Ohio, 44106; $995; Analyses markets for major thermoplastics. Over 20 end-use markets are analysed including construction, transportation, furniture, packaging, appliances, and consumer products.

105; Thermoplastic polyesters; October 1982; Business Communications Co., Inc.; Box 2070 C, Stamford, Connecticut 06906; $950; Growth areas highlighted. Includes markets by resins, structure, changes.

106; Transparent plastics; March 1982; Business Communications Co., Inc.; Box 2070 C, Stamford, Connecticut 06906; $1,500; Analyses markets by type, applications and substitutes, predictions, corporations involved. Latest in acrylics, polycarbonates, styrenics and polyesters.

107; *US Foamed Plastic Markets and Directory*; Annual; Technomic Publishing Co.; 851 New Holland Ave., P.O.Box 3535, Lancaster, Pennsylvania 176104; $35.00; An alphabetical list of 600 urethane foam manufacturers, blowing agents manufacturers, fabricators, processors, laminators, applicators. Polystyrene bead producers, polystyrene foam sheet, film and block producers, processors and fabricators.

108; *US Plastics in Building and Construction*; Periodic; Technomic Publishing Co.; 851 New Holland Ave., P.O. Box 3535, Lancaster, Pennsylvania 17604; $ — unknown; Alphabetical list of 550 manufacturers and contractors of plastic siding, pipe, bath and plumbing fixtures, coatings and paints, wire and cable, flooring, etc.

109; Upgrading thermoplastics; September 1981; Business Communications Co., Inc.; Box 2070 C, Stamford, Connecticut 06906; $1,500; Newest innovations, developments by major resin, technologies, systems, compounding materials, markets, costs, competitive factors, etc.

110; Urethane chemicals (US); August 1979; Frost and Sullivan; 106 Fulton St., New York, New York 10038; $800; Market outlook for urethanes and the major raw materials used in the production of urethane polymers for flexible and rigid foams, elastomers, sealants and surface coatings.
111; Water based polymers; June 1978; Business Communications Co., Inc.; Box 2070 C, Stamford, Connecticut 06906; $725.
112; Water soluble polymers, III; July 1983; Skeist Laboratories, Inc.; 112 Naylon Ave., Livingston, New Jersey 07039; $6,500; Covers technical, economic and marketing information. Major suppliers, users, market shares and capacities. Also industry structure; appraisal of trends, growth rates, new product needs; and government regulations. The changing climate in Washington on environment and energy. Expectations for enhanced oil recovery.

General textbooks

113; *Concise Guide to Plastics*; Periodic; Reinhold Publishing Corp.; 600 Summer Street, Stamford, Connecticut 06901; $ — unknown; Describes the types of properties, forms and basic chemicals from which plastics are made as well as resin manufacture, processing applications by materials and by industries, production and prices. Roster of principal producers and historic review of production. Outlook for plastics use. Identification of trade names.
114; *Markets for Plastics*; Periodic; Reinhold Publishing Corp.; 600 Summer Street, Stamford, Connecticut 06901; $ — unknown; Presents technical information about the science and technology of plastics, books for further reading and lists of organisations and publications serving the plastics industry.
115; *Sourcebook of New Plastics*; Periodic; Reinhold Publishing Corp.; 600 Summer Street, Stamford, Connecticut 06901; $ — unknown; Contains information on 69 leading producers, their executives and products.
116; Trade names dictionary (2nd edn); 1979; Gale Research Co.; Book Tower, Detroit, Michigan 48226; $85.00 (Supplement $83.75); Guide to over 130,000 consumer-oriented trade names, brands, product names, giving the name of the manufacturer, importer or distributor, the address and the source from which the compiler took the data.
117; *US Industrial Directory* (4 Vol); Annual; Cahners Publishing Co.; 275 Washington St., Newton, Massachusetts 02158; $70.00; Information on 40,000 industrial suppliers. Gives addresses,

telephone, product line, employee size category. Supplies catalogues, brochures and related literature.

Plastics industry manufacturers

118; *DuPont Magazine*; E.I. duPont de Nemours and Co., Inc.; Wilmington, 19898; Free; House organ featuring duPont materials and products.
119; *Goodchemco News*; B.F. Goodrich Chemical Group; 6100 Oak Tree Boulevard, Cleveland, Ohio 44131; Free; House organ featuring Goodrich materials and products.
120; *Polysar Progress*; Polysar Ltd., Advertising and Sales Promotion Dept., Sarnia, Ontario N7T 7VI, Canada; Free; House organ featuring Polysar materials and products.
121; *USI News*; U.S.I. Chemicals, National Distillers and Chemical Corp.; 99 Park Ave., New York, New York 10016; Free; House organ featuring U.S.I. materials and products.

South America

122; Brazil '82 plastics and resins outlook and importer profiles; 1982; William D. Gersumsky, Sp Latin America, 301 E. 69th St., New York, New York 10021; $675; Prospects for growth and profitability to 1990. Capacity/demand ratios. Profiles of 150 importers.
123; *Noticiero del Plastico-Elastomeros*; Monthly; Noticiero del Plastico; Alsina 971, 5 Piso Buenos Aires, Argentina; $50/year; trade publication covering plastics and elastomers industry in Argentina.
124; *Pastiguia*; Annual; ASIPLA; Avada. Pedro de Valdivia 1481, Santiago, Chile; $ — unknown; Directory of Chile plastics manufacturers.
125; *Plasticos em Revista*; Monthly; Plasticos em Revista Editora Ltda.; Rua Marques de Itu 95, Sao Paulo, Sp, Brazil; Cr. $2700/year; Trade publication covering plastics industry in Brazil.
126; *Plasticos en Columbia*; Annual; ACO PLASTICOS; CRA. 10a No. 27–27, Interior 137, Of. 901, Bogota, Colombia; $ — unknown; Directory of Colombian plastics companies. Includes production, sales.
127; The Argentine Chemical and Plastics Industry; October 1980; William D. Gersumsky, Sp Latin America; 301 E. 69th St., New York, New York 10021; $750; Data includes producers, production, imports/exports, end-use patterns.

Table 19.1 Index of listings

Subject	*Source*
ABS	44, 47, 53, 66
Acrylic	40, 47, 53, 71, 106
Acquisition	90
Additives	2, 64, 65, 73
Adhesives	10, 11, 61, 84, 95
Alloys	66, 93, 96,
Applications	10, 12, 20, 26, 27, 43, 45, 50–52, 82, 87, 90, 96, 106, 107, 113
Argentina	123, 127
Automotive	37, 71
Bags	68, 80, 102,
Benzene	63
Blends	96, 93
Blow moulding	107, 43
Blowing agents	107
Brazil	125
Bottles	67
Building	71, 108
Business machines	79
Butadiene	63
Butylene	53, 63
Canada	12, 13
Capacities	2, 112,
Chemicals	14, 15, 16, 17, 33, 38, 55, 62, 63
Chile	124
Coatings	110, 95, 84, 108
Co-extrusion	93
Colombia	126
Competition	48, 54, 71, 75, 77, 87, 96, 98, 109
Composites	42, 89
Compounding	98, 109
Construction	71, 108
Consultants	11
Consumers	82, 101, 103, 104, 116
Consumption	70, 77, 78, 88, 97, 101
Containers	43, 86, 115
Converters	59, 101
Cost	3, 42, 61, 63, 71, 76, 91, 109
Design	22, 23
Directories	8, 13, 31, 32, 38, 46, 107, 117
Economics	42, 55, 57, 81, 83, 112
Elastomers	28, 53, 66, 84, 110
End-use	5, 18, 38, 36, 40, 41, 58, 60, 67, 65, 69, 70, 75, 84, 87, 95, 96, 97, 99, 104, 112, 127,
Energy	62, 63, 112
Engineering resins	47, 66
Environment	60, 112
Equipment	6, 25, 61, 70, 93, 94
Ethylene	63
Expense	3
Exports	39, 54, 76, 127
Extenders	51

Subject	*Source*
Extrusion	93, 95
Fabrication	93, 99, 100, 107
Fabricators	99, 107
Feedstocks	2, 62, 63
Fillers	51
Film	58, 59, 60, 61, 68, 95, 107
Financial	3, 86
Finished products	2, 91
Foam	69, 79, 84, 85, 110, 107
Forecasts	42, 45, 50, 53, 62, 64, 65, 72, 79, 86
Forming	74, 103
Furniture	104
Future	48, 52, 55
Georgia	35
Government — regulations,	32, 40, 44, 60, 67, 84, 86, 112
Growth rates	5, 40, 60, 77, 79, 86, 90, 93, 105, 112
Health	60
High temperature	52
Imports	33, 39, 54, 76, 116, 122, 127
Intermediates	33, 38, 95
International trade	54, 76
Inventory	3
Japan	83
Labour survey	4
Laboratories	11
Machinery	2, 21, 25, 70, 72, 94, 97
Manufacturers	11, 28, 30, 37, 43, 97, 107, 108, 116, 124
Markets	2, 9, 14–16, 29, 42, 44, 45, 49, 53, 57–59, 64, 66, 67-69, 75, 80, 85, 88, 89, 92, 94, 97, 100, 104, 105, 106, 109
Materials	8, 10, 12, 18, 20, 21, 23, 26, 27, 37, 48, 49, 50, 55, 75, 76, 77, 84, 85, 87, 89, 91, 92, 93, 96, 103, 109, 110, 113, 118, 119, 120, 121
Membership (SPE, SPI)	7, 8
Metal	82
Mexico	1, 19
Mississippi	34
Netting	61
New processes	97, 102, 103,
New products	40, 41, 43, 60, 83, 90, 109, 112
North Carolina	31
Nylon	47, 89
Operating ratios	3
Packaging	68, 77, 87, 104
Paper	80
Paperboard	81
Phenolic	44
Pigments	33, 38
Pipe	92, 108
Plant locations	46
Plasticizers	95, 38
Polyarylates	83
Polycarbonate	47, 71, 106
Polyesters, thermoplastic	47, 71, 95, 105, 106

CHAPTER TWENTY

Japan

I. OMAE

The enduring post-war renaissance of Japan as a major industrial nation has been one of the most remarkable phenomena of our times. Significant growth has been achieved across all sectors of industry, not least in plastics.

Production of the principal plastics materials in Japan is shown in Table 20.1. Production for 1985 was 9.23 million tonnes, an increase of 4% over 1984, of which thermoplastics amounted to 7.53 and thermosetting resins 1.65 million tonnes. For the first time, polyethylene terephthalate (0.31) and polybutylene terephthalate (0.02 million tonnes) had specific entries in the thermoplastic figures, otherwise the total production of both thermoplastics and thermosets was roughly the same for 1984 and 1985. Polymers achieving significant increases in production over 1984 included polyvinyl alcohol (122%), fluorine resins (125%), polycarbonate (116%) and polyacetal (114%).

Recent uses found for polyvinyl alcohol include materials for paper diapers, sanitary napkins, soil water retention materials, freshness retention materials for fruit and vegetables and water retention materials for medical applications such as poultices and disposable body warmers etc.

Fluorine resins are widely used in electronics, car and aircraft parts, precision machinery components and in medical apparatus. Polycarbonate and polyacetal engineering plastics have been used for optical electronics e.g. for compact discs, and car parts.

However, some plastics showed a decline in production, e.g. melamine resin (91.9%), epoxy resin (89.4%), polyethylene (90.1%), and ethylene-vinyl acetate copolymer (81.4%).

Table 20.1. Production of plastics in Japan (million tonnes) (adapted from Matsuzaki (1986), courtesy of *Plastic Age*)

Plastic	1981	1982	1983	1984	1985	'85/'84	1986 Jan-June
Phenolics	0.294	0.281	0.303	0.334	0.327	97.8	0.155
Urea resin	0.504	0.487	0.511	0.487	0.470	96.5	0.226
Melamines	0.125	0.119	0.120	0.117	0.107	91.9	0.053
Unsatd.p'ester	0.182	0.190	0.192	0.193	0.202	104.5	0.096
Alkyds	0.130	0.137	0.134	0.145	0.144	99.5	0.068
Epoxy	0.069	0.068	0.082	0.102	0.091	89.4	0.044
Silicone	0.044	0.047	0.058	0.063	0.068	108.1	0.039
Urethane foam	0.209	0.211	0.222	0.228	0.244	106.8	0.123
Thermosets	1.558	1.540	1.623	1.670	1.653	99.0	0.804
LDPE	1.033	1.031	0.963	1.257	1.119	89.0	0.541
HDPE	0.638	0.643	0.688	0.841	0.785	93.3	0.389
EVA copolymer[b,c]	nd[a]	nd	0.122	0.153	0.124	81.0	0.058
Polystyrene	0.803	0.798	0.857	0.984	1.065	106.4	0.499
AS resin	0.081	0.083	0.098	0.108	0.112	103.6	0.058
ABS	0.300	0.310	0.355	0.411	0.422	102.6	0.221
Polypropylene	0.959	0.941	1.062	1.271	1.304	102.5	0.660
Polybutene	0.023	0.028	0.029	0.031	0.031	100.1	0.017
Petroleum resin	0.080	0.085	0.088	0.107	0.107	100.3	0.050
Methacrylate	0.104	0.112	0.128	0.141	0.147	104.2	0.076
Polyvinyl alc.	0.095	0.093	0.098	0.108	0.133	122.8	0.090
PVC	1.129	1.218	1.420	1.503	1.550	103.0	0.744
Polyvinylidenes	0.032	0.033	0.033	0.037	0.039	105.0	0.017
Polyamide	0.069	0.079	0.089	0.107	0.112	104.7	0.056
Fluoroplastic	0.005	0.005	0.006	0.010	0.013	125.1	0.007
Polycarbonate	0.033	0.036	0.037	0.043	0.050	115.7	0.033
Polyacetal[b]	nd	nd	0.063	0.075	0.086	114.2	0.054
PET[c]	nd	nd	nd	nd	0.314	nd	0.167
PBTP[c]	nd	nd	nd	nd	0.019	nd	0.013
Thermoplastics	5.384	5.495	6.138	7.189	7.530	104.7	3.747
Other plastics	0.097	0.100	0.051	0.055	0.048	87.5	0.023
Total	7.038	7.135	7.812	8.914	9.232	103.6	4.573

[a] nd = no data. [b] Polyacetal is classified as 'Other plastics' and EVA copolymer as LDPE prior to 1983. [c] EVA = Ethylene/vinyl acetate; PET = polyethylene terephthalate; PBTP = polybutylene terephthalate.

World production of plastics materials in 1985 is estimated to be 76.48 million tonnes of which Japan produced 9.23 million tonnes, (12.1%) second only to the USA which produced 22 million tonnes, (29%).

Engineering plastics such as fluorine resins, polycarbonate, polyacetal, polyethylene terephthalate (PET) and polybutylene terephthalate (PBTP) amounted to 0.482 million tonnes in 1985 which was larger than that of the USA which produced 0.133 million tonnes of polycarbonate and 0.049 tonnes of polyacetal, Matsuzaki (1986).

Production of advanced engineering plastics such as modified polyphenylene oxide/polyphenylene ether, polyphenylene sulphide, polymethylpentene, polysulphone, polyethersulphone, polyimide, polyarylate etc. is shown in Table 20.2.

Table 20.2. Production and demand — special engineering plastics (from Matsuzaki (1986), courtesy of ***Plastic Age*****)**

	1983	1984	1985	
Modifed PPO/PPE	37,800	45,100	46,900	demand
Polyphenylene sulphide (PPS)	1,800	2,800	3,700	demand
Polymethylpentene	nd	nd	3,000	demand
Polysulphone	360	470	540	demand
Polyethersulphone	130	170	200	demand
Polyimide	220	nd	500	demand
Polyarylate (U polymer)	500	700	800	prodn.

Tables 20.1 and 20.2 demonstrate therefore that Japan has been engaged in active research and development of the main engineering and advanced types of plastics. In addition, recent research has been focused on separation membranes, high absorption polymers, conductive polymers, medical applications for polymers, liquid crystal polymers, optical fibres, photosensitive polymers, polymer alloys, natural polymers, composites, photoconductive polymers etc. Conductive polymers include polyacetylenes, polypyrroles, polyanilines, polythiophenes, PPS, and polyphenylene vinylenes.

Polymers for medical applications include blood-compatible polymers, e.g. segmented polyurethanes, polyfluoroethylene and polysilicones, biodegradable polymers and methyl methacrylate/2-hydroxyethyl methacrylate copolymers for contact lenses.

Optical fibre polymers include polymethyl methacrylate, polystyrene and polycarbonate core polymers, surrounded by a sheath of vinylidene fluoride/tetrafluoroethylene copolymer or fluoroalkyl methacrylate polymer.

Periodicals

Periodicals covering polymers in Japan are published by the Society of Polymer Science, Japan, associations concerned with polymer science, publishers, newspaper publishing companies, etc.

Periodicals published by the Society of Polymer Science, Japan contain academic articles, whereas periodicals published by the associations concerned with polymers and the majority of the remaining periodical publishers present a wide range of technical articles.

THE SOCIETY OF POLYMER SCIENCE, JAPAN

Periodicals published by the Society include: *Kobunshi (High Polymer Japan* 1952–vol.37, 1988), *Kobunshi Ronbunshu (Journal of Polymer Science* 1944–vol.45, 1988), *Polymer Journal* (1970–vol.20, 1988). *Kobunshi* is the organ of the Society of Polymer Science, Japan, carrying papers on polymer science in Japanese, while on the other hand, *Polymer Journal* carries articles on polymer science in English.

ASSOCIATIONS CONCERNED WITH POLYMER SCIENCE

Periodicals covering polymer science include:

Bulletin of the Chemical Society of Japan (1928–vol.61, 1988) (English edition published by the Society)

Chemical Letters (1972–) (English edition published by the Chemical Society of Japan)

Convertech (1973–vol.16, 1988) (*Paper Film and Foil*, published by the Converting Technical Institute)

Gosei Jushi (1955–vol.34, 1988) (*Plastics*, published by the Japan Society of Plastics Technology)

Kagaku To Kogyo (1948–vol.41, 1988) (*Chemistry and Industry*, published by the Chemical Society of Japan)

Kagaku To Kogyo (Osaka) (1926–vol.62, 1988) (*Science and Industry*, published by Osaka Koken Kyokai)

Kamipa Gikyoshi (1947–vol.42, 1988) (*Japan TAPPI Journal*, published by the Japan Technical Association of the Pulp and Paper Industry)

Kyoka Plastics (1955–vol.34, 1988) (*Reinforced plastics*, published by the Japan Reinforced Plastics Society)

Netsukokasei Jushi (1955–vol.9, 1988) (*Journal of Thermosetting Plastics, Japan*, published by the Japan Thermosetting Plastics Industry Association)

Nihon Gomu Kyokaishi (1929–vol.61, 1988) (*Journal of the Rubber Society of Japan*, published by the Society of the Rubber Industry of Japan)

Nihon Rheology Gakkaishi (1973–vol.16, 1988) (*Journal of Rheology, Japan*, published by the Society of Rheology, Japan)

Nihon Sanshi Gakkaishi (1930–vol.27, 1988) (*Journal of Sericultural Science of Japan*, published by the Japanese Society of Sericultural Science)

Nihon Setchaku Kyokaishi (1965–vol.24, 1988) (*Journal of the Adhesion Society of Japan*, published by the Society)

Nippon Kagaku Kaishi (1972–) (*Journal of the Chemical Society, Japan*, published by the Society)

Petrotech (1978–vol.11, 1988) (published by the Japan Petroleum Institute)
Sekyu Gakkaishi (1958–vol.31, 1988) (*Journal of the Japan Petroleum Institute*, published by the Institute)
Sen-I Gakkaishi (1945–vol.44, 1988) (*Journal of the Society of Fibre Science and Technology, Japan*, published by the Society)
Sen-I Kikai Gakkaishi (1948–vol.41, 1988) (*Journal of the Textile Machinery Society of Japan*, published by the Association of the Textile Machinery Society of Japan)
Shikizai Kyokaishi (1928–vol.61, 1988) (*Japan Society of Colour Material*, published by the Society)
Yuki Gosei Kyokaishi (1943–vol.46, 1988) (*Journal of Synthetic Organic Chemistry, Japan*, published by the Society of Synthetic Organic Chemistry, Japan)
Zairyo (1952–vol.37, 1988) (*Journal of the Society of Material Science, Japan*, published by the Society)

OTHER POLYMER RELATED PERIODICALS

Included below are some of the titles produced by book publishers, news publishing companies, research institutes, chemical companies, etc.
Enbi To Polymer (1961–vol.28, 1988) (*Vinyls and Polymers*, published by the Institute of Polymer Industry Inc.)
Japan Plastics Age (1963–vol.25, June 1987) (English edition, published by Plastics Age Co. Ltd., publication suspended)
JETI (1953–vol.36, 1988) (*Japan Energy and Technology Intelligence*, published by Saiwai Shobo Co. Ltd.)
Kino Zairyo (1981–vol.8, 1988) (*Function and Materials*, published by CMC Co. Ltd.)
Kobunshi Kako (1952–vol.37, 1988) (*Polymer Applications*, published by Kobunshi Kankokai Co. Ltd.)
Kogyo Zairyo (1953–vol.36, 1988) (*Engineering Materials*, published by the Daily Industrial Newspaper Co. Ltd.)
Plastic Seikei Gijutsu (1985–vol.5, 1988) (*Plastic Mold Technology*, published by Shiguma Co. Ltd.)
Plastics (1950–vol.39, 1988) (Published by Kogyo Chosaka Publishing Co. Ltd.)
Plastics Age (1955–vol.34, 1988) (Published by Plastics Age Co. Ltd.)
Polyfile (1964–vol.25, 1988) (Information on polymer technology and polymer industry published by Taiseisha Co. Ltd.)
Polymer Digest (1949–vol.40, 1988) (Published by Rubber Digest Co. Ltd.)

Setchaku (1957–vol.32, 1988) (*Adhesion and Sealing*, published by Kobunshi Kankokai Co. Ltd.)
Speciality Chemicals (1977–vol.12, 1988) (Published by Tekumato Co. Ltd.)
Toso Gijutsu (1962–vol.27, 1988) (*Japan Finish*, published by Riko Co. Ltd.)
Toso To Toryo (1929 –) (*Finishes and Paints*, published by Toryo Shuppan Co. Ltd.)

Abstracts

The major abstracts publication is *Current Bibliography on Science and Technology* (1927–vol.62, 1988) published by the Japan Information Center for Science and Technology (JICST).

About 500,000 abstracts are produced each year from 12,000 journal titles, reports and proceedings. Abstracts in the chemistry and chemical engineering fields total about 150,000 per year of which about 30,000 are of Japanese articles. 6,000 out of the total are on polymer related topics and about 1,000 of these are from Japanese language originals.

The most recent abstracts, 1975 to date, are conveniently acessed using the JICST data base on the JOIS online retrieval system discussed below.

Other abstracts journals include *Kaigai Kobunshi Kenkyu* (1952–vol.34, 1988) and *Kagaku Shoho* (1973–). The former contains abstracts of polymer articles from foreign country sources and is published by the Society of Polymer Science, Japan. The latter contains abstracts from about 120 Japanese chemical periodical titles and is published by Kagaku Joho Kyokai (Japan Association for International Chemical Information — JAICI).

Patents

Japanese patents started in 1885 with 425 patent applications, increasing to the enormous annual total of more than 320,000 in 1986. Utility models started in 1905 reaching 204,000 per annum in 1986. Other patents in Japan concern designs and trademarks. Overall, the total number of applications in 1986 in Japan amounted to 746,000 as shown in Table 20.3. The total number of applications for patents and utility models in 1984 was 486,984, about 44% of the world total in 1984 of 1,107,138 (Annual Report of Patent Office, 1987).

Patents are classified by the International Patent Classification (IPC). A breakdown of Japanese patent applications by IPC subclass in the plastics and polymer fields is given in Table 20.4.

Table 20.3. Number of patent applications

	1982	1983	1984	1985	1986
Patents	237,513	254,956	284,767	302,995	320,089
Utility models	202,706	205,243	202,181	204,815	204,210
Designs	59,390	57,618	54,683	55,237	52,636
Trademarks	139,198	150,318	161,882	161,546	168,890
Total	638,807	668,135	703,513	724,593	745,825

Table 20.4. Japanese patent applications by IPC subclass (Annual Report of Patent Office (1987), Vol 39, 101)

IPC		1981	1982	1983	1984	1985
B29	Working of plastics; working of substances in a plastic state in general	3296	3390	3606	3881	4039
B31	Making paper articles; working paper	215	177	170	191	172
B32	Layered products	1629	1588	1796	1844	1799
B65	Conveying; packing; storing; handling thin or filamentary material	4965	5302	5806	6561	6865
CO7	Organic chemistry	6613	6823	6666	7085	7477
CO8	Organic macromolecular compounds, their prepn, or chemical work-up and compsns. based thereon	5549	5810	6505	7413	7775
CO9	Dyes; paints; polishes; natural resins; adhesives; miscellaneous compsns.; miscellaneous applications of materials	2812	2945	3061	3174	3090
C10	Petroleum, gas, coke industries technical gases containing carbon monoxide; fuels; lubricants; peat	1373	1385	1336	1065	1134
C11	Animal or vegetable oils, fats, fatty substances or waxes; fatty acids therefrom; detergents; candles	342	396	441	392	540
C13	Sugar industry	49	40	41	40	27
C14	Skins; hides; pelts; leather	24	25	24	14	26
DO1	Natural or artificial threads or fibres; spinning	812	975	968	1105	1081
DO2	Yarns; mechanical finishing of yarns or ropes; warping or beaming	371	370	407	401	359
DO6	Treatment of textiles or the like; laundering; flexible materials not otherwise provided for	1767	1822	1938	2073	1816
D21	Paper-making; production of cellulose	520	457	435	464	443

Databases

Online information systems originating in Japan include JOIS, PATOLIS, NEEDS-IR, DIALINE, Infostream, HINET (Technosearch, Titlesearch, HIASK, Asahi Shinbun Kiji data base and Nikkei Sangyo Shibun Kiji data base), ID-IR, NOCS, COSMOS, TONETS, KINO-DIAL, NEEDS-TS, NIKKEI TELECOM, BRANDY (trademarks), TSR (BIGS/FINES), MARK III, Assist,

Cornet (architecture), ACE, CALL SERVICES, LEX/DB (law), MARVIS (local area) and PPDS (physical properties of chemical compounds), etc.

NEEDS-TS, NIKKEI-TELECOM, TSR (BIGS/FINES). MARK III CALL SERVICES, DATA MAX, COSMOS, etc. are mainly business data base systems while NEEDS-IR, HINET, ACE and NK-MEDIA in JOIS are mostly newspaper-oriented data base systems. IRSPAN in Infostream and ID-IR are data base systems containing infrared spectra data.

NOCS, TONETS, BOOK in Assist and KINODIAL are data bases on recently published books and JAPAN/MARC in DIALINE is the data base system for books in the National Diet Library collection.

Of all these systems, JOIS and PATOLIS are the most important systems to use for searching for polymer information. Databases in JOIS are shown in Table 20.5

JICST, JCLEARING, JPUBLIC, JICST E, NK-MEDIA, JAFIC and OSAKA-UE in JOIS are Japanese data bases. JICST covers all scientific fields and may be usefully searched with the Japanese control term *jugotai* (polymerized materials) to retrieve polymer science information.

KJP in PATOLIS is the Japanese patents data base from which information on Japan Kokai Koho (applications open to public

Table 20.5. JOIS data bases–June 1987

Database	*Range*	*No.items (annual)*	*No.items (total)*	*Subject*
JICST	1975–	540,000	5,200,000	Science
JCLEARING	1981–		50,000	On-going research
JMEDICINE	1981–	220,000	460,000	Medicine in Japan
JPUBLIC	1983–	4,000	12,000	Official reports
JICST E	1985–	160,000	330,000	Medicine in Japan (English language)
NK-MEDIA	1983–	20,000	70,000	Industrial news
JAFIC	1985–	4,000	6,000	Foods in Japan
OSAKA-UE	1970–	1,000	4,000	Urban technology
MEDLINE	1972–	300,000	4,100,000	Medicine
TOXLINE	1977–	200,000	1,540,000	Toxicology
CANCERLIT	1963–	60,000	590,000	Cancer literature
CA SEARCH	1977–	480,000	4,670,000	Chemistry
BIOSIS	1979–	470,000	3,000,000	Biology
CAB	1979–	190,000	1,330,000	Agriculture
NTIS	1981–	70,000	440,000	Technology
INSPEC	1981–	210,000	1,270,000	Physics/electronics
FSTA	1981–	20,000	120,000	Foods
EMBASE	1980–	260,000	1,880,000	Pharmaceuticals
COAL	1978–86		110,000	Coal
INIS	1972–	90,000	1,090,000	Nuclear science
IRRD	1981–	15,000	90,000	Road
INFOTERRA			5,400	Environment (on-going)

inspection), published eighteen months after filing at the Japan Patent office or after the priority date claimed on the application.

Abstracts of Japanese patents in selected subject fields have been translated into English since 1976. However, fields selected for translation have changed three times, the first period covered 1976–1979 (excluding Japanese private patent applications), the second period covered 1980–1984 and the third from 1985 to the present date, as shown in Table 20.6.

Recently the JAPIO data base has been made available in ORBIT, EasyNet, or Iquest in the CompuServe system (Simmons, 1986). The JAPIO data base consists of two parts, one being a data base with translated abstracts of the patents applied for by Japanese inventors and the second introduced from the INPADOC data base which consists mainly of the patent data applied for by foreign inventors. In contrast to the first part of the JAPIO where search words are taken from the abstract, there is no abstract attached in INPADOC, with the result that only the title words and IPC codes are available for subject searching. JAPIO is a large data base as may be seen from Table 20.7.

Table 20.6. Selected fields for translation into English of Japan Kokai Koho

M Field			*C Field*		*E Field*		*P Field*
B21	EO2	F16J	AO1N	CO7	HO1F	HO3	GO1
B22	FO1	F16K	A23	CO8	HO1J	HO4	GO2
B23	FO2	F16N	A61K	CO9	HO1L	HO5G	GO3
B24	FO3	F17	BO1	C10	HO1M		GO4
B27N	FO4	F23	BO3	C12	HO1P		GO5
B29	F15	F24	BO4	C21	HO1Q		GO6
B30	F16C	F28	BO5	C22	HO1S		G11
B41	F16D		BO9	C23	HO2K		
B60	F16F		CO1	C25	HO2M		
B62D	F16G		CO2	C30	HO2N		
B65G	F16H		CO3	DO1	HO2P		
B65H				DO6N			

M = General and mechanical; C = chemical; E = electrical; P = physical.

Table 20.7. Annual increments to the JAPIO data base

Publication year	*Nos. added*
1976	40,796
1977	156,263
1978	149,013
1979	163,577
1980	166,553
1981	169,103
1982	211,843
1983	224,282
1984	231,582
1985	262,662
1986	294,825
Total	2,070,499

Conclusions

Although polymer information is readily available from international sources e.g. from *Chemical Abstracts* on systems such as Dialog, BRS, ORBIT, STN International, information on polymer developments in Japan e.g. on JICST in JOIS, KJP in PATOLIS (or in English from JAPIO in ORBIT or EasyNet), is the most comprehensive and perhaps more easily obtainable from these sources by ordering the original journal articles and patents online.

There is much information appearing in the sources mentioned which is not available anywhere else. Japan will continue to be an important source of information for the forseeable future notwithstanding the language difficulties for Western countries and it is hoped that this outline of the major sources will provide some useful leads for searchers.

References

Annual Report of Patent Office (1987) **39**, (101), 4669.

Matsuzaki, S. (1986). Trend of plastics industry in Japan, *Plastics Age*, December, 111–142.

Simmons, E.S. (1986). 'Japio — Japanese patent applications on-line, *Online*, July, 5´ -58.

CHAPTER TWENTY-ONE

The Pacific Basin and India

M. B. McALLISTER

Any study of the information sources in plastics and polymers in the region of the Pacific Basin must take into account three factors. The first is the overwhelming dominance in the region of Japan, both industrially and intellectually. The sheer size of the Japanese output of papers and patents makes the contribution of other countries in the area pale into insignificance. Secondly, the region is the largest rubber-producing area in the world, with Malaysia being the world's major rubber producer; this fact undoubtedly has the effect of suppressing research and development in synthetic rubbers, plastics and polymers. Thirdly, a number of the countries are ex-colonies or developing countries and the major chemical industries are subsidiaries of larger multinationals whose research efforts have, in the past, been concentrated in their parent companies in Europe or the USA, although this is slowly changing.

Regional sources

Directories

Two directories dealing specifically with the Pacific region should be noted, although neither confines its scope to plastics and polymers. *Pacific Research Centres: a Directory of Organisations in Science, Technology, Agriculture and Medicine* (Longman, 1986) is one of the most recent publications in the *Longman's Reference on Research* series. Each entry lists the parent organisation with subsidiary research centres and contains information on senior staff, activities, budget and publications. The countries covered are Australia, Brunei, People's Republic of China, Hong Kong, Indonesia, Japan, Korea, Malaysia, New Zealand, Papua New Guinea, Philippines, Singapore, Taiwan, Thailand and Vietnam.

An older work, although equally comprehensive is *Technological Research and Development Institutions in Asia and the Pacific* (Economic and Social Commission for Asia and the Pacific of the United Nations, 1982).

Journals

It is an indication of the growing interest in the region from overseas that a new journal has appeared, published in the UK but dealing specifically with Asia. This is *Polymers and Rubber Asia: News, Information and Comment.* (London, S.K.C. Communication Services, 1985–). A number of the established international journals contain columns and regular features on research and industry in the Pacific region. *Plastics and Rubber International* (London, Plastic and Rubber Institute, 1976–) is a well known example. *Polymer News* (New York, Gordon and Breach, 1976–) has a column entitled Polymers in Asia and North Africa, another on Polymers in Australia, both contributed by local experts; in addition its series on Centers of Polymer Research often features Asian and Pacific research institutes. *Chemical and Engineering News* (Washington, D.C., 1923–) in its annual June survey of foreign chemical industries publishes useful comparative statistics for Korea, Australia, China and Taiwan.

Australia

Research

Australia has a vigorous research effort in plastics and polymers and the Polymer Section of the Royal Australian Chemical Institute (RACI) is a particularly active professional group which produces the useful directory *Australian Polymer Science and Engineering* (RACI, Melbourne, 2nd edn, 1986) with details of staff and programmes of academic, government and industrial laboratories.

A directory in the broader area of materials science but containing much information of interest is *Directory of Materials*

Research and Development Capabilities and Facilities in Australia (Department of Science, 1986). This is the first edition of a directory which it is hoped will be updated regularly. It provides detailed descriptions of laboratories and is well indexed. Much of Australian research is co-ordinated through the Commonwealth Scientific and Industrial Research Organisation (CSIRO). T. Ermers of their Information Resources Unit has edited *Scientific and Technical Research Centres in Australia* (Melbourne, 1984). This is also available for online searching on the AUSTRALIS system through CSIRONET.

Most of the work in polymers and plastics in Australia is published in the major international journals. Some coverage is given by *Australian Journal of Chemistry* (CSIRO, 1948–) and *Chemistry in Australia* (RACI, 1934–) but these are both general chemistry journals and the proportion of polymer-related topics is low. The Polymer Section of the RACI hold an annual *Australian Polymer Symposium* and also organize regular workshops and seminars on specific topics in polymer science such as biomedical polymers. A new initiative is the launch of a monograph series *Australian Polymer Science*; the first volume by Z.H. Stachurski is *Engineering Science of Polymeric Materials* (RACI, 1987).

Industry

A detailed discussion of the Australian plastics industry is available in the *Report of the Industries Assistance Commission Inquiry into the Chemicals and Plastics Industries* (Canberra, AGPS, 1986). The industry is almost entirely devoted to fulfilling domestic demand. The industry association is the Plastics Institute of Australia (PIA) which has both corporate and professional membership. It is responsible for a number of useful publications. *Plastics News* (Melbourne, 1950–) is a monthly trade magazine with news of companies, techniques and trade shows; it also contains specialist trade directories in selected issues. *The Plastics Institute of Australia Trade Directory* (Melbourne, PIA) is an annual listing of companies in the field. *Who Supplies Which Plastics for What?* (PIA, Raw Materials and Additives Division, 1983) is a comprehensive, well indexed directory of suppliers of raw materials. The Plastics and Rubber Institute has an Australian section open to professionals in the industry; it organizes workshops and seminars and reports of its activities may be found in *Plastics and Rubber International*.

Statistics of production may be obtained from the PIA and also from the Australian Bureau of Statistics. The Standards Association of Australia (SAA) is responsible for setting standards relating to the industry and details can be found in their annual *Catalogue of Publications* (Sydney, SAA). At present there are over 100 standards relating to polymers and plastics. The Australian Patent Office is a member of the World Patent Organisation and Australian patents are reported in INPADOC and in the Derwent publications and data bases.

China

One of the most exciting developments in the Pacific region in recent years has been the emergence of the People's Republic of China from a long period of isolation. The last few years have seen an increase in the information available about China and its science and technology.

Research

Chinese research establishments feature in *Pacific Research Centres* already referred to in the introduction. Another, somewhat dated work is Susan Swannack-Nunn's *Directory of Scientific Research Institutes in the People's Republic of China* (Washington, D.C. National Council for US-China Trade, 1977); Vol II, *Chemicals, Construction*, contains a wealth of information regarding the structure of scientific research in China and includes detailed descriptions of research institutes, their staff and significant achievements. A useful up-to-date source of information regarding academic societies, institutes and people in China is contained in the annual *China Directory in Pinyin and Chinese* (Tokyo, Radiopress Inc.).

Two Chinese language journals on the subject of plastics and polymers are abstracted by *Chemical Abstracts* which makes them more accessible to Western readers. They are *Gaofenzi Tongxun [Polymer Communications]* (Beijing, China Int. Book Trading Corp., 1956–) and *Suliao [Plastics]* (Beijing, China Int. Book Trading Corp., 1972–). Both journals have summaries in English. Mention should be made here of an English language abstracting journal covering Chinese journals called *China Science and Technology Abstracts* (Hong Kong, International Information Service, 1980–). It is in three parts and *Series II, Chemistry, Earth Sciences, Energy Sources* and *Series III, Industrial Technology* both contain references to polymer and plastics research. The abstracts are sketchy and in some cases nonexistent but it is a useful alerting tool.

Industry

An industrial process journal indexed in *Chemical Abstracts* is *Hecheng Xiangjiao Gongye [Synthetic Rubber Industry]* (Lanzhou, 1978–). Information regarding manufacturers, including plastics, polymers and rubber manufacturers may be found in *China Directory of Industry and Commerce and Economic Annual* (Boston, Science Books International). This large reference work has a detailed index and for each plant lists the Chinese name and address, English name and address, name of directory, number of employees and short description of products. Statistics of industrial production including the chemical industries are available from the State Statistical Bureau of the People's Republic of China, Beijing.

In addition, annual statistics and general informative articles may be found in the *People's Republic of China Yearbook* (Evergreen Publishing Co., Hong Kong, in conjunction with Xinhua Publishing House, Beijing).

Another indication of China's awakening industrial effort is the appearance of a data base *Chinese Patent Abstracts in English*, produced by the Patent Documentation Center of the People's Republic of China. This file is available on both the DIALOG and ORBIT systems and contains patent numbers, assignees, inventors and a short English abstract. Approximately 15% of the patents relate to polymers or plastics.

India

Research

As would be expected from a large, civilised country with a scientific tradition dating back for many centuries and an abundance of raw materials, India has a thriving research base in academic, government and industrial areas. Polymer science began in India as early as 1920, however the science has boomed since the 1950s. There are a number of centres of polymer research including the Indian Institute of Technology at New Delhi and the National Chemical Laboratory at Poona. Indian scientists have no lack of journals in which to publish their work. The *Journal of the Institution of Chemists (India)* (Calcutta, 1929–), the *Indian Journal of Chemistry. Section B, Organic and Medicinal Chemistry* (New Delhi, Council for Scientific and Industrial Research, 1976–), the *Journal of the Indian Chemical Society* (Calcutta, 1924–) all publish papers covering the whole field of chemistry including polymers. There is also the specialist *Journal of Polymer Materials* (Calcutta, Oxford and IBH Publishing, 1984–) which is sponsored by the Indian Institute of Technology.

Industry

The development of both the rubber and plastics industry in India is well documented in an outstanding technical encyclopedia edited and published by the Council of Scientific and Industrial Research (CSIR). This is *The Wealth of India* (New Delhi, 1947–1980), a work in two series, each of nine volumes. Series One deals with *Raw Materials*, including rubber. Series Two is titled *Industrial Products* and Vol VII, PL-SH, covers the rubber and plastics industries in some depth. Each chapter consists of a technical description, history of the industry and present position, with many diagrams, photographs and statistics. All is supported by detailed bibliographic references which are themselves a valuable guide to the literature. The volume in question was published in 1971 and a second edition is in progress.

In a country as large as India trade and professional associations abound and each state has its own quota too numerous to mention in a work of this scope. There are a number of national industry associations, The All India Plastic Manufacturers Association which publishes the trade journal *Plastics News* (Bombay) and the All India Rubber Industries Association which is responsible for *Rubber India* (Bombay, 1949–). The Indian Rubber Manufacturers Research Association sponsors a regular Rubber Conference. Other important journals are *Rubber Board Bulletin* (Kottayam, 1951–), *Plastics, Rubber and Leather Industries Journal* (Baroda, 1962–), *Indian Rubber and Plastics Age* (Bombay, 1966–) and *Rubber and Plastics Digest* (New Delhi, 1965–) and *Popular Plastics* (Bombay, 1955–) which is a colourful and informative trade journal. All are in English and most are indexed by either *Chemical Abstracts* or *RAPRA*.

Information about plastics, polymer and rubber manufacturers in India may be obtained from the above journals but an excellent compilation of industry information is contained in *Kothari's Economic and Industrial Guide of India* (Madras, Kothari and Sons, 34th edn, 1983). It is divided into industries, each section has an introduction giving the history of the industry and present situation, including statistics. Detailed information is provided for each company, including address, personnel, products and share capital. Unfortunately the useful annual *Times of India Directory and Yearbook including Who's Who* ceased publication in 1984; it was a useful source for industry and professional associations, government departments and statistics. Statistical information is available in the annual *Statistical Abstract of India* (New Delhi, Central Statistical Organisation). The Indian Standards Institution has published a large number of standards on rubber, polymers and plastics.

Korea

Research

South Korea has fast growing manufacturing strength which is backed up by academic and industrial research. This is no less true of the plastics and polymer industry than of, for instance, electrical equipment. The co-ordinating body for the application of scientific research to industry is the Korea Institute for Industrial Economics and Technology; it publishes, in addition to journals in Korean, a most useful abstracting service in English. This journal, *Korean Scientific Abstracts* (Seoul, 1968–) is bimonthly and contains lengthy, informative abstracts from the major Korean scientific and industrial journals.

A number of Korean journals publish polymer and plastic related research, among them are *Taehan Hwahakhoe Chi [Journal of the Korean Chemical Society]* (Seoul, 1963–) and *Hwahak Konghak [Journal of the Korean Institute of Chemical Engineers]* (Seoul, 1963–). There is also one journal devoted to the subject called *Pollimo[Polymer]* (Seoul, Polymer Society of Korea, 1977–). All of these journals are indexed by *Chemical Abstracts* and also by *Korean Scientific Abstracts*.

Industry

Some relevant industry associations are The Korean Institute of the Rubber Industry which publishes the quarterly journal *Komu Hakhoechi* (Seoul,1980–), the Korean Tire Industrial Association which is responsible for *Taio Gomu [Tire and Rubber]* (Seoul, 1973–) and the Korean Plastic Industry Cooperative.

The names of individual manufacturers may be obtained from one of a number of directories, the most comprehensive of which is *Korean Exporters and Manufacturers Annual* (Korean Trade Promotion Corporation, Seoul) which has a very detailed subject index and for each company gives number of employees, address and name of president. Another useful source is *Business Korea Yearbook* which is produced by the publishers of the English language periodical *Business Korea*; it contains the names and addresses of trade associations as well as individual companies.

Industry statistics are available from the Bureau of Statistics of the Korean Economic Planning Board, Seoul. Standards are the responsibility of the Bureau of Standards, Seoul.

New Zealand

Research

Most scientific research in New Zealand has a bias towards agriculture and earth sciences. The Department of Scientific and Industrial Research (DSIR) publishes *New Zealand Scientific Abstracts* (Wellington, 1980–) which indexes not only material published in New Zealand journals but also material published overseas by New Zealand researchers. Although not comprehensive it is a good overview of research in New Zealand and polymers and plastics feature rarely. The New Zealand Institute of Chemistry which publishes *Chemistry in New Zealand* (Wellington, 1936–) has a polymer section which organizes regular seminars.

Industry

In the manufacturing area there has been a dramatic growth in an industry which barely existed 20 years ago and in the last two years production has doubled. This is in the main due to the implementation of the recommendations of the *Industries Development Commission Report on the New Zealand Plastics Industry*, (Wellington, 1981). The major industry body is the Plastics Institute of New Zealand (PINZ) which produces an annual directory, the *New Zealand Plastics Industry Directory* (Trade Publications Ltd., Auckland). This is a comprehensive trade directory containing a useful Trade Names Index.

Statistics on the industry are published by the New Zealand Bureau of Statistics and the Standards Association of New Zealand is responsible for standards.

Taiwan

Although Taiwan is a major producer of polymers and plastics, there is no defined body of literature dealing with the subject and information sources are hard to identify.

Research

There are two bodies in Taiwan which promote science and industry. One is government sponsored, the other independent but non-profit-making. The National Science Council (NSC) administered by the government has a number of research centres, some of which are involved in polymer research. It publishes, in English, a journal *Proceedings, Series A and B*.

The Industrial Technology Research Institute, which is independent, runs the Union Chemical Laboratories at Hsinchu, which as well as performing research in polymer related areas, provides a consulting service through its Plastic Processing Service Group.

The Science and Technology Information Center of the National Science Council publishes a *Directory on Sci-Tech R & D Institutions in R.O.C.* (Taipei, 1985). This book lists about 500 laboratories but its usefulness is limited by the lack of a subject index.

Industry

A useful introduction to Taiwanese industry is the wide-ranging English language publication *Industry of Free China* (Taipei, Council for Economic Planning and Development, 1954–). This contains in-depth articles on many Taiwanese industries. The Taiwan Plastics Industry Association publishes a journal *Su Chiao T'ung Hsun [Plastics Correspondence]* (Taipei) which is indexed in *Chemical Abstracts*. Statistics on production are published by the Directorate-General of Budget Accounting and Statistics and yearly tables are available in the useful *Statistical Yearbook of the Republic of China* (Taipei).

Other countries

In addition to the countries covered above, there are small but important rubber, polymer and plastics industries in such countries as Sri Lanka, Hong Kong and Singapore. These countries all have local branches of the Plastics and Rubber Institute and the journal *Plastics and Rubber International* carries regular reports from them. Singapore has an active polymer research group at the Singapore Polytechnic which publishes the annual *Plastichem* (Singapore, 1969–).

CHAPTER TWENTY-TWO

Translations

E. INGLIS

Those of us who live in the English-speaking world are perhaps fortunate in that for most purposes we may not have to burden ourselves with an extensive knowledge of foreign languages. After all, a vast amount of information on polymers (and on practically every other subject) is written in English anyway: by British, Americans, Australians and by many other nationalities who realize that their work will reach a wider public if written in one of the major culture-languages, English being the usual choice.

People such as Finns or Hungarians are in a different situation from ourselves since few people other than Finns speak Finnish, for example — even in Finland, Swedish is accepted as the alternative national language. Hence, to gain access to the considerable amount of literature on polymers and related fields, it is advisable to read more than one language, especially if your native language happens to be spoken by a world minority.

There are many journals throughout the world dealing with rubber and/or plastics: *Kauchuk i Rezina*, *Plasticheskie Massy*, *Vysokomolekulyarnye Soedineniya* (Russian); *Plasty a Kaučuk* (Czechoslovakian); *Polimery, Tworzywa Wielkocząsteczkowe* (Polish); *Materiale Plastice* (Rumanian); *Polimeri* (Yugoslavian); *Nippon Gomu Kyokaishi* and *Kobunshi Ronbunshu* (Japanese); *Muovi Viesti* (Finnish); *Müanyag és Gumi* (Hungarian); *Plaste und Kautschuk* (German); *Kunststof en Rubber* (Dutch); *Materie Plastiche ed Elastomeri* (Italian); *Plastiques Modernes et Elastomères* (French); *Plast Nordica* (Scandinavian); and *Revista de Plasticos Modernos* (Spanish).

In addition to these of course there are also many journals covering special applications such as tyres (or tires), packaging, industrial rubber products, etc., which most information centres no longer have the money to buy regularly, though some, such as the British Library Document Supply Centre at Boston Spa, do aim at a fairly comprehensive coverage of the world's journal literature.

In addition to these monthly journals mentioned above there is also a good deal of foreign language information available in books, standards and patents, which have been mentioned in other chapters. These are frequently referred to in the bibliographies of journal articles, though quite frequently difficulties are encountered, such as the patent numbers being wrong, authors' names garbled, wrong year of publication, etc. Regrettably, nothing is ever quite plain sailing in the world of information.

Some foreign journals carry what purport to be summaries of the papers they contain, written in other languages for the benefit of English, French, German, etc. speakers, but these are often somewhat suspect as they are rarely written by linguistic experts and occasionally give quite the wrong impression of what the paper is about. The reason for this seems to be that when it comes to translation everybody wants to do things 'on the cheap'.

Ideally of course the translator should have an expert command of both the source-language (the language one translates from) and the target-language (the language one translates into). If either of these is more to be preferred than the other, I would say that it is advisable for the translator to be familiar with his target-language, and preferably to be a native speaker of it.

Formerly many firms involved in the polymer industry had their own experienced teams of translators, but this type of information resource has tended not to be renewed, owing to financial pressures, and most translations nowadays are done by translation agencies, some of which have been fortunate enough to engage the services of redundant translators who have many years of experience in the job.

The best value is gained from translation agencies if they can provide someone who is not only a good linguist but is thoroughly conversant with the subject matter. This is a tall order, since it implies that the translator should know all about polymer chemistry, physics, biomedical applications, chemical analysis, tyres, packaging, etc. Few people have spent their lives becoming experts in all these things, and in addition have found time to become fluent in German, French, Hungarian or whatever the language or range of languages offered might be. This imposes a compromise, in which translators of polymer literature are either people who have a linguistic training and who have also become interested in translation of polymer literature, or who are trained scientists or engineers who have acquired a knowledge of a language or language. It would be foolhardy to say that either of these two types is inherently better than the other, except perhaps that acquiring a difficult language at the end of one's scientific career with a view to editing or translating is a formidable though not impossible undertaking. What is required for translation of course, is a good reading knowledge of the language, not good pronunciation and verbal fluency, which are the requirements of an interpreter rather than a translator.

The difficulty of a translator who is not conversant with the study of polymers and their applications is that he simply does not know what quite a lot of words mean. There are a number of specialist dictionaries which he may consult, such as: *A Russian-English Dictionary for the Rubber and Plastics Industries* (M. Lambert, RAPRA Technology Ltd., MacLaren and Sons); *Wittfoht's Plastics Technical Dictionary (Kunststoff-Technisches Worterbuch)* (Hanser); *Elsevier's Rubber Dictionary* (10 languages); *Dorian's Dictionary of Science and Technology* (German-English, Elsevier); German-English *Dictionary of Plastics Technology* (M.S. Welling, Pentech Press); *Dictionary of Polymer Technology* (in Japanese, with English glossary equivalents, Taiseisha); *Czech-English Technical Dictionary* (eds. B. Kloudová and V. Stackeova, SNTL); *Czech-English Technical Dictionary* (eds. J. Feigl and E. Klinger, SNTL); *Polish-English Technological Dictionary* (eds. S.C. Zerni and M. Skrzynska, Wydawnictwe Naukowe-Techniczne), and Callaham's *Russian-English Chemical and Technological Dictionary* (Wiley and Sons). The long-term solution to this problem is to take an intelligent interest in what one is translating and this is where an 'in-house' translator has an advantage, since he is able to pick the brains of experts and consult books in the information/library area.

The disadvantage of specialised dictionaries is that they consume a lot of space with somewhat redundant information, for example that 'polietilen' in Russian means polyethylene, 'polivinilkhlorid' means polyvinyl chloride, and other such transparently obvious concepts.

They may however, omit to tell you other things which you might like to know, and which would be found in a more general dictionary. It is wise to remember that specialised dictionaries are not intended as substitutes for the comprehensive general dictionary, and are simply an additional aid. Many translators prefer to keep and update their own specialised glossaries, mistrusting published dictionaries on principle, and improving their knowledge and learning from their mistakes.

The difficulty of an editor examining a translation from a

language with which he is unfamiliar is that some unscrupulous translators may provide a text which is perfectly plausible but is not a complete translation, i.e. it may have sections where passages were missed out because the translator found them difficult, or could not find the appropriate words in his dictionary. An honest translator will admit this, and suggest that the translation be looked at by an expert. Even then, the expert may not have the imagination to realize that for example 'an inverted refrigerator' or 'upside-down icebox' is what he would normally term a reflux condenser, or he may be puzzled because he is unaware that polyether and polyester are expressed by the same word in Russian. Unfortunately there is no way of avoiding the kind of problem caused by unfamiliarity with a language, except by acquiring a thorough knowledge of it.

However, a little knowledge is better than no knowledge at all, and those who just want to get the general idea of what a text is about rather than having a full translation of it may be content with a smattering of the language. This kind of need is catered for by a course run every year by Sheffield's Japanese Department, where in a few weeks the student, often with no knowledge at all, is taught to 'decode' Japanese texts with the aid of a dictionary and Dr. Janacek's *Japanese Grammar Dictionary*. This method involves a programmed examination of the text in much the same way as a translating computer would look at it, and can lead to a greater interest in Japanese, and even to a desire to learn the language properly!

Everyone who has tried to acquire information by translating it will be aware that translating information on polymers from e.g. Russian is not the same sort of exercise as passing one's Finals in Russian Translation. There may be misprints which you are not told about in advance, the author may have a cumbersome repetitive style, he may be under the impression that he has expressed elegantly what he wanted to say when in fact he hasn't, several lines may have been transposed by the printers without anyone noticing, etc.

The translation journal *International Polymer Science and Technology*, (RAPRA Technology Ltd.) includes translations of papers mainly from East European journals, with some Japanese, i.e. the sort of areas where translation expertise may not be so readily accessible to the polymer information researcher. Translations are chosen by readers' selection from a list of titles of papers and authors. *Polymer Science USSR* (Pergamon) is a cover-to-cover translation of the Soviet journal *Vysokomolekulyarnye Soedineniya A (High-Molecular Compounds, Part A)*. The main problem in setting up such a translation service or product is, of course, gaining permission to translate and publish. Some Japanese journals such as *Kobunshi Ronbunshu* publish their own English-language versions, but even in the original Japanese version many of the diagrams and Tables have captions in English.

A great deal of work and effort is currently being expended on the development of computer translation programs to take the donkey-work out of translating. The producers of such systems, such as *Automated Language Processing Systems* (Data General, Solihull), which already exist for a number of languages, normally insist that these are translation aids rather than translators, but the programs do provide rough texts which can be readily edited, as on a word-processor. The editor can make the more difficult decisions on language problems which the computer was unable to resolve, such as when the sentence in the source-language is ambiguous or ungrammatical (once again the human element creates the difficulty), and the use of such translating machines, allowing rapid translation of vast amounts of text, is undoubtedly the way things will go in the future.

Index

Note: a separate, more detailed index to Chapter 19 appears on p 276–278.